Introduction to Electromagnetism

Introduction to
Electromagnetism
From Coulomb to Maxwell

Second Edition

Martin J. N. Sibley

CRC Press
Taylor & Francis Group
Boca Raton London New York

CRC Press is an imprint of the
Taylor & Francis Group, an **informa** business

Second edition published 2021
by CRC Press
6000 Broken Sound Parkway NW, Suite 300, Boca Raton, FL 33487-2742

and by CRC Press
2 Park Square, Milton Park, Abingdon, Oxon, OX14 4RN

First edition published by Butterworth-Heinemann 1995

CRC Press is an imprint of Taylor & Francis Group, LLC

Library of Congress Cataloging-in-Publication Data
Names: Sibley, M. J. N. (Martin J. N.) author.
Title: Introduction to electromagnetism : from Coulomb to Maxwell / Martin J. N. Sibley.
Description: Second edition. | Boca Raton : CRC Press, 2021. |
Includes bibliographical references and index. |
Summary: "This edition aims to expand the on the first edition and take the reader through to the wave equation on coaxial cable and free-space by using Maxwell's equations. The new chapters will include time varying signals and fundamentals of Maxwell's equation"— Provided by publisher.
Identifiers: LCCN 2020049401 (print) | LCCN 2020049402 (ebook) |
ISBN 9780367460563 (hardback) | ISBN 9780367462703 (ebook) |
Subjects: LCSH: Electromagnetism.
Classification: LCC QC760 .S48 2021 (print) | LCC QC760 (ebook) |
DDC 537—dc23
LC record available at https://lccn.loc.gov/2020049401
LC ebook record available at https://lccn.loc.gov/2020049402

ISBN: 978-0-367-46056-3 (hbk)
ISBN: 978-0-367-71187-0 (pbk)
ISBN: 978-0-367-46270-3 (ebk)

Typeset in Times
by codeMantra

To my family.

Contents

Preface

In editing the first edition to produce this text, I have tried to keep the ethos of making electromagnetism accessible to as many people as I can. With this in mind, I have kept the fundamental nature of the book the same. I have changed and updated the examples used where necessary.

The major change is the addition of sections dealing with Maxwell's equations and their application. This is an area of electromagnetism that is often considered difficult. By taking a slightly unconventional route to some ideas, I hope that readers will gain an insight into the basics of time-varying electromagnetic fields.

The discussion of the Joint European Torus (JET) is still present, but be aware that the Next European Torus (NET) is due to be operational in 2022, so watch the Internet.

Author

Dr. Martin J. N. Sibley studied at the then-Polytechnic of Huddersfield (Huddersfield, UK) obtaining a BSc (Hons) in Electrical Engineering in 1981. He stayed at Huddersfield to do a PhD, conferred in 1984, researching preamplifiers for optical receivers. This work was sponsored by British Telecom Research Laboratories (BTRL) and resulted in the first preamplifier to be fabricated in IC form. A BTRL Research Fellowship followed, examining the practical design of a Digital Pulse Position Modulation (DPPM) coder and decoder. Theoretical predictions showed that DPPM could outperform Pulse Code Modulation (PCM), and practical results obtained using the experimental system confirmed the predictions. This work was practically based as was most of his subsequent research.

In 1986, Dr. Sibley joined the academic staff at the, now, University of Huddersfield. His main research area was optical communications, in particular the practical implementation of Visible Light Communication (VLC) using high-power LEDs, and coding schemes for highly dispersive optical links such as plastic optical fibre. He has acted as consultant to several blue-chip companies.

Dr. Sibley is now retired having published over 80 conference and proceedings papers as well as five text books in the fields of electromagnetism, optical communications and telecommunications.

1 Introduction

This book is concerned with the study of electrostatic, electromagnetic and electro-conductive fields – sometimes referred to as field theory or, more simply, electromagnetism. A knowledge of this subject can help us to explain why a circuit refuses to behave as designed, why components sometimes break down and what happens in high-frequency circuits. In studying this area, life is made a lot easier if we can think in three dimensions. This is usually a case of drawing adequate diagrams and practicing.

Readers used to circuit theory may wonder why they should study such a discipline. Well, field theory is the study of some of the fundamental laws of Nature. Indeed, electromagnetism was the first theory to unite the sciences of electricity and magnetism. The search is now on to find a Grand Unified Theory that unites all the basic forces of Nature, and that should be of interest to us.

As we progress with our studies, we will meet some names that have become famous in the field of electrical engineering. Some of these people have had units named after them, and so will be more familiar than others. Before we begin our studies in earnest, let us take a moment to pay our respect to some of the researchers who contributed to electrical engineering as we know it today.

1.1 HISTORICAL BACKGROUND

Electromagnetic field theory is really the result of the union of three distinct sciences. The oldest of these is electrostatics, which was first studied by the Greeks. They discovered that if they rubbed certain substances, they were able to attract lighter bodies to them. One of these substances was amber, whose Greek name is *electron* – this is where we get the name 'electricity'. It was in 1785 that French physicist, Charles Augustin de Coulomb (1736–1806), showed that electrically charged materials sometimes attract and sometimes repel each other. This was the first indication that there were two types of charge – positive and negative.

In the late 1700s, two Italians were working on the new science of current electricity. One, Luigi Galvani (1737–1798), was a physiologist and physician who thought that animal tissues generate electricity. Although he was later proved wrong, his experiments stimulated Count Alessandro Volta (1745–1827) to invent the first electric battery in 1800. Most of the early experiments in current electricity were performed on frog's legs – this was a result of Galvani's work.

Later, a favourite party trick was to get a group of people to hold hands and then connect them to a voltaic cell (a battery). The cell produced quite a large voltage, which then caused current to flow through the guests. This made them jump uncontrollably! It wasn't until 1833 that the British experimenter Michael Faraday (1791–1867) showed that the current electricity of Volta and Galvani was the same as the electrostatic electricity of Coulomb. Rather than linking these two phenomena, it was shown that the current and electrostatic electricity were one and the same thing.

(Faraday's contribution is all the more remarkable when it is realized that his theories were formulated by direct experimentation and not by manipulating mathematics!)

Although the ancient Greeks also knew about magnetism in the form of lodestone, the Chinese invented the magnetic compass, and in 1600, William Gilbert of Gloucester laid down some fundamentals. However, it was not until 1785 that Coulomb formulated his law relating the strengths of two magnetic poles to the force between them. Magnetism may have been laid to rest here if it wasn't for the Danish physicist Hans Christian Oersted (1777–1851). It was Oersted who demonstrated to a group of students that a current-carrying wire produces a magnetic field. This was the first sign that electricity and magnetism could be interlinked. This link was strengthened in 1831 by the work of Faraday who showed that a changing magnetic field could induce a current into a wire. It was a French physicist André Marie Ampère who first formulated the idea that the field of a permanent magnet could be due to currents in the material. (We now accept that electrons orbiting the nucleus constitute a current, and this produces the magnetic field.)

We owe our present view of 'field theory' to Faraday who performed many experiments on electricity and magnetism. Although Faraday preferred to work without mathematics, he did introduce the idea of fields in free-space. This greatly influenced later workers, and it was in the mid-1800s that the British physicist James Clerk Maxwell (1831–1879) formalized Faraday's results using mathematics. Among other things, Maxwell was able to predict the existence of electromagnetic waves. This work inspired others in the field, such as Oliver Heaviside (1850–1925) who worked on the first transatlantic telegraph cable as well as predicting the existence of the ionosphere.

The rest, as they say, is history. Due to the work of the German physicist Heinrich Rudolf Hertz (1857–1894) and the Italian engineer Guglielmo Marconi (1874–1937), we are now able to communicate over vast distances. We can also use electrical machinery to make our lives more comfortable. In fact, we owe our current way of life to the hard work of a number of researchers who continually questioned and experimented, carefully recording their results and observations.

1.2 ATOMIC STRUCTURE

When we learn to drive a car, we do not necessarily need to know exactly how the car works. However, if we do understand how the engine works and why the wheels turn, it can help us to be better drivers. A similar situation occurs with electricity and magnetism – when we use electricity and magnetism, we seldom have to worry about exactly how the effects are produced. However, it can make us better engineers if we have an adequate model of what electricity and magnetism are. This is where we have to study the structure of the atom.

Figure 1.1 shows the basic structure of the simplest atom, the hydrogen atom. This atom has one electron that orbits the nucleus containing a single proton. The charge on the electron is equal and opposite to the charge on the proton and has the value

$$e = -1.6 \times 10^{-19} \text{ C} \tag{1.1}$$

with units of coulomb, symbol C.

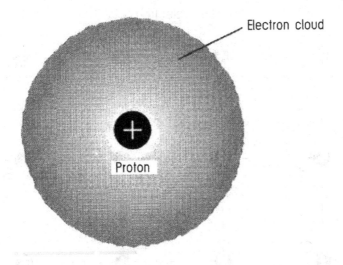

FIGURE 1.1   Basic structure of a hydrogen atom.

More complex materials, such as amber for instance, have many atoms held in a crystalline structure. If we rub amber, the friction removes electrons, so leaving the material positively charged. This is the basis of electrostatic electricity. In some materials, the electrons are very tightly bound to the nucleus and considerable energy must be expended to remove an electron. These are insulators.

In a metal atom, the electron in the outermost orbit is not tightly bound to the nucleus. When a number of metal atoms are close to each other, they form a crystalline structure in which these outermost electrons are free to move around; see Figure 1.2. Now, metals are usually electrically neutral with the number of electrons exactly balancing the number of protons. If we connect a source of electrons to the metal, injected electrons will travel through the lattice. As like charges repel, these electrons force the free electrons away from them. The net effect is to produce a disturbance that travels down the metal. The rate of flow of charge is the electric current. We should note that the mass of a proton is 1837 times the mass of an electron; thus, conduction in metals is by electron flow.

Let us now turn our attention to magnetism. As we will see in Chapter 3, a current generates a magnetic field. In an atom, we can regard the motion of electrons around a nucleus as constituting a current. Thus, there will be a magnetic field. In most materials, the electron orbit is completely random, and so there is no perceptible magnetic field. However, in some materials, e.g. iron, the electrons can travel in the same general direction. Thus, each atom becomes a permanent magnet. When these atoms are part of a crystalline structure, their magnetic fields are randomly distributed, and there are negligible external effects. However, if we subject the material to an external magnetic field, the atomic magnets align themselves to the field. When we remove the field, some of these atomic magnets stay in their new positions, so producing a permanent magnet.

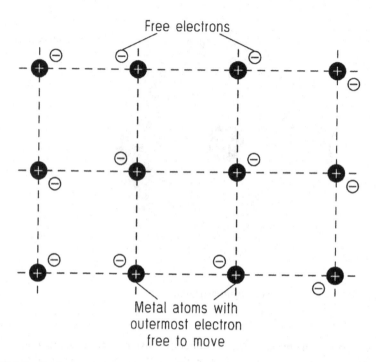

FIGURE 1.2   Basic crystalline structure of a metal.

This section has introduced us to some basic ideas about electricity and magnetism. Although this discussion has been very simplistic in form, the models we developed will be useful in later chapters. We will now consider vector notation, and briefly examine some coordinate systems.

1.3   VECTORS AND COORDINATE SYSTEMS

When we use a thermometer, we read the temperature off a graduated scale. The temperature of a body is independent of direction (it is simply measured at a certain point), and so it is known as a scalar quantity. Scalar quantities are those that have no direction associated with them.

If we push an object, we have to exert a force on it. This force has direction associated with it – we could push the object to the left, to the right or in any direction we choose. The force is a vector quantity because it has magnitude and direction.

At this point, we could launch into a discussion of vector theory – addition, multiplication, etc. Unfortunately this would complicate matters, and mask the underlying ideas. Instead, we will avoid vector algebra in favour of discussion and reasoning. In spite of this, Figure 1.3 shows the standard Cartesian, spherical and cylindrical systems that we will use as we progress with our studies. (We will use unit vectors in most of the text, however. This is to help readers get used to vector notation, which will aid future studies.)

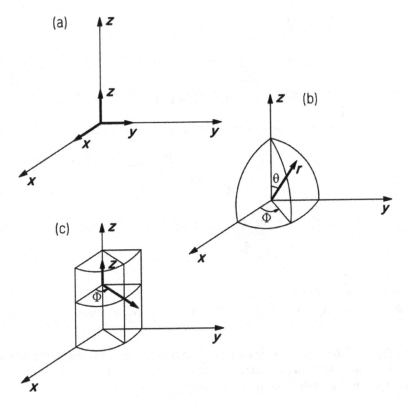

FIGURE 1.3   (a) The standard Cartesian coordinate set, (b) the spherical coordinate set and (c) the cylindrical coordinate set.

1.4   LINE, SURFACE AND VOLUME INTEGRALS

This book assumes that readers are familiar with basic integration and differentiation. However, we will come across many instances where we need the line, surface and volume integrals. As most readers will not be very familiar with these integrals, this section deals with their definition and application.

Figure 1.4a shows a line of length l. Let us consider a small incremental section of the line of length dy. Now, the left-hand end of the line is at the origin of a Cartesian coordinate set, and the line lies along the y-axis. When we do the line integral, we effectively add together the lengths of the incremental section as we move it along the line. This is represented by

$$\text{length} = \int_0^l dy$$

$$= |y|_0^l$$

$$= l$$

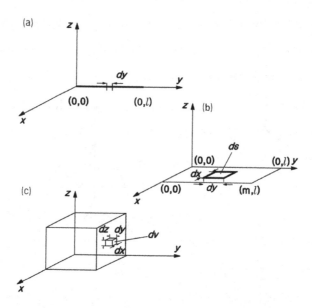

FIGURE 1.4 (a) Line integral, (b) surface integral and (c) volume integral.

So, the line integral merely gives us the length of the line. As regards the surface integral, Figure 1.4b shows a square lying in the xy-plane. Let us take a small incremental section of area ds. This area is given by

$$ds = dxdy$$

Now, when we perform a surface integral, we effectively move ds across the whole of the square. We can split this into two parts: we can integrate with respect to y to give a line of thickness dx and length l, and then integrate with respect to x to give the total area. So,

$$\text{area} = \int_0^m \left(\int_0^l dy \right) dx$$

$$= \int_0^m l\,dx$$

$$= ml$$

The volume integral follows a similar procedure: we consider a small incremental volume and integrate with respect to x, y and z. This is shown in Figure 1.4c. So, the volume integral is

$$\text{volume} = \int_0^n \left(\int_0^m \left\{ \int_0^l dy \right\} dx \right) dz$$

$$= \int_0^n \left(\int_0^m l\,dx \right) dz$$

$$= \int_0^n lm\,dz$$

$$= lmn$$

Although we have confined ourselves to a Cartesian coordinate set, we could have considered the spherical or cylindrical sets. With these sets, the same basic principal applies – consider a small incremental section, and then integrate with respect to the relevant coordinates. Problems 1.1–1.5 should provide readers with practice!

1.5   STRUCTURE OF THIS BOOK

The book is essentially in two parts. The first is concerned with fundamentals and linking a circuits point of view to a fields approach. It culminates in a discussion of the various field equations that are relevant to a deeper discussion of radio waves and electromagnetic fields in general. The second part of the book is concerned with the solution of the wave equations for coaxial cable and free-space. These wave equations predict the propagation of waves with distance and time. We are familiar with such phenomena in the form of radar signals that travel in space and in time. A study of wave propagation in a planar optical waveguide is used as an application of Maxwell's equations.

There is some rather complex mathematics in the later sections of this book, which can be omitted at a first reading. The important result is that there are waves that vary with time and distance – travelling waves. These are all around us in the form of radio waves, and it is a concept that readers would do well to understand.

2 Electrostatic Fields

Most of us are familiar with the phenomenon of electrostatic discharge: lightning strikes, sparks from nylon clothing and sparks from nylon carpet. It may be thought that the study of static electric fields has little to offer the electrical engineer. After all, we are taught that electrons flow in conducting materials, and so why should we concern ourselves with the study of static charges? However, as we shall see later in this chapter, electrostatics introduces several ideas that will be very helpful when we consider capacitors and, ultimately, transmission lines.

2.1 COULOMB'S LAW

As we have seen in Chapter 1, electronic charge comes in two forms: negative charge from an electron and positive charge from a proton. In both cases, a single isolated charge has a charge of 1.6×10^{-19} Coulomb. If there are two charges close to each other, they tend to repel each other if the charges are alike or attract each other if they are dissimilar. Thus, we can say that these charges exert a force on each other.

Charles Augustin de Coulomb (1736–1806) determined by direct experimental observation that the force between two charges is proportional to the product of the two charges and inversely proportional to the square of the distance between them. In terms of the SI units, the force between two charges, a vector quantity, is given by

$$F = \frac{q_1 q_2}{4\pi \varepsilon r^2} r \qquad (2.1)$$

where
F is the force between the charges (N)
q_1 and q_2 are the magnitudes of the two charges (C)
ε is a material constant (F m^{-1})
r is the distance between the charges (m)
and r is a unit vector acting in the direction of the line joining the two charges
– the radial unit vector

This is Coulomb's law. The force, as given by Equation (2.1), is positive (i.e. repulsive) if the charges are alike, and negative (i.e. attractive) if the charges are dissimilar (see Figure 2.1). As Equation (2.1) shows, the force between the charges is inversely dependent on a material constant, ε, the permittivity. Good insulators have very high values of permittivity, typically ten times that of air for glass and so the electrostatic force is correspondingly smaller.

If no material separates the charges, i.e., if they are in a vacuum, the permittivity has the lowest possible value of 8.854×10^{-12} or $1/36\pi \times 10^{-9}$ F m^{-1}. (These rather obscure values result from the adoption of the SI units.) As permittivity has such a low value, it is more usual to normalize the permittivity of a material to

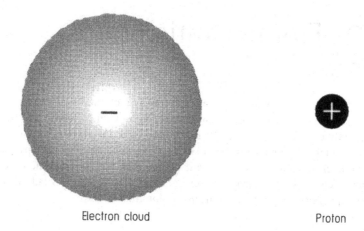

Electron cloud   Proton

FIGURE 2.1   Two separate charges in free-space.

that of free-space. This normalized permittivity is commonly known as the relative permittivity, ε_r, given by

$$\varepsilon_r = \frac{\varepsilon}{\varepsilon_o} \qquad (2.2)$$

With this form of normalization, ε_r ranges from 1.0006 for air to 5–10 for a good insulator such as glass.

Before we consider an example, it is worth examining Coulomb's law in greater detail. One of the first things we should note is that Coulomb's law incorporates the inverse square law, i.e., the force is inversely proportional to the square of the distance between the charges. This relationship is more commonly found when considering gravitational fields; however, we will meet it again when we study magnetic fields. Another point worth noting is the presence of $4\pi r^2$ in the denominator of Equation (2.1). This is simply the surface area of a sphere and we will see why this is so when we consider electric flux in the next section.

Example 2.1

Determine the force between two identical charges, of magnitude 10 pC, separated by a distance of 1 mm situated in free-space. What is the force if the separation is reduced to 1 μm?

Solution

The force between two charges is given by Coulomb's law, Equation (2.1), as

$$F = \frac{q_1 q_2}{4\pi \varepsilon r^2} r$$

Thus,

$$F = \frac{10 \times 10^{-12} \times 10 \times 10^{-12}}{4 \times \pi \times 8.854 \times 10^{-12} \times \left(1 \times 10^{-3}\right)^{2}} r$$

$$= 0.9 \times 10^{-6} r \text{ N (repulsive)}$$

If we reduce the separation to 1 μm, the force increases to 0.9 N.

By way of contrast, if we consider a hydrogen atom, the single electron is in orbit around the single proton at a minimum distance of 5.3×10^{-11} m. This gives an attractive force of 8.2×10^{-8} N. Thus, we can see that the electrostatic force in an atom is very small.

In this section, we have seen that charges exert a force on each other. This force is repulsive if the charges are alike and attractive if the charges are unlike. This effect raises the question: how does one charge 'know' that the other is present? To answer this, we will introduce the idea of electric flux.

2.2 ELECTRIC FLUX AND ELECTRIC FLUX DENSITY

One definition of flux is that it is the flow of material from one place to another. Some familiar examples of flow are water flowing out of a tap or spring, air flowing from areas of high pressure to low pressure and audio waves flowing outward from a source of disturbance. In general, we can say that flux flows away from a source and towards a sink.

If we adapt this to electrostatics, we can say that a positive charge is a source of electric flux, and a negative charge acts as a sink. We must exercise extreme caution here. Nothing physically flows out of positive charges – a charge does not run out of electric flux! What we are doing is adapting the general definition of flux, so that we can visualize what is happening. If we consider isolated point charges, we can draw a diagram as in Figure 2.2. (A point charge is simply a physically small charge or collection of charges. This raises the question of how small is small? The answer lies with relative sizes. Relative to the distance between the Earth and the Sun, the height of Mount Everest is insignificant. Similarly, we can regard a collection of individual charges, arranged in a 10-nm diameter sphere, as a point charge when viewed from 10 m away.)

Now, what happens to the distribution of electric flux if we bring two positive charges together? As the charges are both sources of electric flux, the fluxes repel each other to produce the distribution shown in Figure 2.3. One of the main things to note from this diagram is the distortion of the lines of flux in the space between the charges. This causes the force of repulsion between the two charges, in agreement with Coulomb's law.

If we now return to Coulomb's law, we can rewrite it as

$$F = \frac{q_1}{4\pi r^2} \frac{1}{\varepsilon} q_2 r \tag{2.3}$$

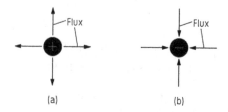

FIGURE 2.2 Flux radiation from isolated point charges: (a) a positive charge and (b) a negative charge.

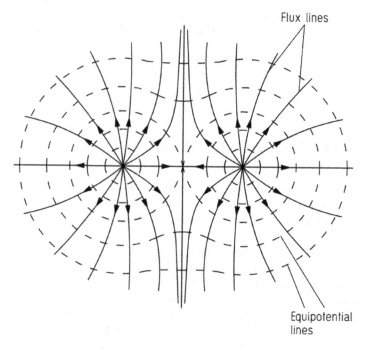

FIGURE 2.3 Distribution of flux due to two positive point charges in close proximity.

The first term in Equation (2.3) consists of the electronic charge, q_1, divided by the surface area of a sphere, $4\pi r^2$. Thus, $q_1/4\pi r^2$ has units of C m^{-2} and would appear to be a surface density of some sort – the flux density. To explain this, we must use Gauss' law (Karl Friedrich Gauss, 1777–1855) which states that the flux through any closed surface is equal to the charge enclosed by that surface.

Figure 2.4 shows an imaginary spherical surface surrounding an isolated point charge. Application of Gauss' law shows that the flux, ψ, radiating outwards in all directions has a value of q_1 – the amount of charge enclosed by the sphere. The area of the Gaussian surface is simply that of a sphere, i.e., a surface area of $4\pi r^2$. Thus, we get a flux density, D, of

$$D = \frac{q_1}{4\pi r^2}r \tag{2.4}$$

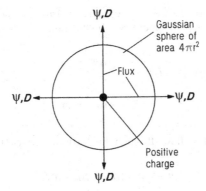

FIGURE 2.4 Relating to the definition of flux density.

As Figure 2.4 shows, the flux density is a vector quantity, i.e., it has direction and magnitude. Specifically, **D** has the same direction as the flux – **away** from the charge for a positive charge and **towards** the charge for a negative charge. Convention dictates that flux radiating away from a charge is positive, whereas the opposite is true for flux going towards a body.

Example 2.2

1. Determine the flux radiating from a positive point charge of magnitude 100 pC.
2. What is the flux density at a distance of 10 mm from the charge?
3. Determine the flux that flows through an area of 200 mm² on the surface of a 1-m radius Gaussian sphere.
4. Repeat (3) if a negative charge of the same magnitude replaces the positive charge.

Solution

1. Application of Gauss' law shows that the flux from the 100 pC charge is simply the magnitude of the charge. Thus,

$$\psi = 100\,pC$$

2. At a radius of 10 mm, the total flux is still 100 pC. However, the surface area of the sphere is $4\pi r^2$ where r is the radius of the sphere. So, as the radius of the Gaussian sphere is 10 mm, we get a flux density of

$$D = \frac{100 \times 10^{-12}}{4\pi\left(10 \times 10^{-3}\right)^2}\,r$$

$$= 7.96 \times 10^{-8}\,r\,C\,m^{-1}\ \text{in a radial direction}$$

3. We now need to find the flux through a 200 mm² area on the surface of a 1-m radius Gaussian sphere. So, the flux density at this radius is

$$D = \frac{100 \times 10^{-12}}{4\pi l^2} r$$

$$= 7.96 \times 10^{-12} r \text{ cm}^{-1}$$

As this flux density flows through the 200 mm² surface, the flux is

$$\Psi = 7.96 \times 10^{-12} \times 200 \times 10^{-6}$$

$$= 1.59 \times 10^{-15} \text{C}$$

4. We now replace the positive charge by a negative one of the same magnitude. As the charge is numerically the same, the magnitudes of all the quantities will be the same. However, as the charge is negative, we have to put a minus sign in front of the answers. So, the flux from the charge is

$$\psi = -100 \, \text{pC}$$

the flux density at 10 mm is

$$D = -7.96 \times 10^{-12} r \, \text{C m}^{-1}$$

and the flux through the 200 mm² surface is

$$\psi = -1.59 \times 10^{-15} \, \text{C}$$

This section has shown us that positive electric charges radiate flux, whereas negative charges attract flux to them. This model enables us to draw field plots such as that in Figure 2.3. Although such plots can help us in visualizing the field surrounding a point charge, we are usually more concerned with the force on a charge due to the presence of a fixed charge. This is where the idea of an electric field becomes useful.

2.3   THE ELECTRIC FIELD AND ELECTRIC FIELD STRENGTH

As we saw in the previous section, we can write Coulomb's law as

$$F = \frac{q_1}{4\pi r^2} \frac{1}{\varepsilon} q_2 r$$

If we use the definition of electric flux density given by Equation (2.4), we can write

$$F = \frac{D}{\varepsilon} q_2 r \qquad (2.5)$$

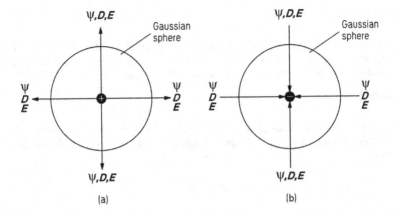

FIGURE 2.5 Flux, flux density and electric field strength in (a) a positive charge field and (b) a negative charge field.

In Equation (2.5), we can see that the force on the charge q_2 is directly proportional to the factor D/ε. This factor has units of newton per coulomb, i.e., N C^{-1}. Thus, we can regard this quantity as the force on a unit of charge. To emphasize this more, we introduce a new parameter known as the electric field strength, E, defined as

$$E = \frac{D}{\varepsilon} \qquad (2.6)$$

We can now write Coulomb's law as

$$F = q_2 E$$

or, more generally,

$$F = qE \qquad (2.7)$$

From this equation, we can see that the force is directly dependent on the electric field strength, also known as the electric field intensity. We should note that, as E is directly proportional to D, it is also a vector quantity.

The field strength is an important parameter in that it introduces us to the idea of a force field. Figure 2.5a shows an isolated point charge at the centre of a Gaussian sphere. As this figure shows, electric flux, ψ, radiates outwards from the charge. Also shown are the electric flux density and electric field strength vectors.

Let us now introduce a positive test charge of 1 C. This charge will experience a repulsive force acting in a radial direction – the direction of the E field. As this test charge is 1 C, the magnitude of the force will also be the magnitude of the E field. Thus, the lines of flux are also the lines of force emanating from the charge. A similar situation arises with a negative charge: Figure 2.5b. So, we can say that a force field surrounds each charge, and that the field is repulsive if the charges are alike, and attractive if the charges are dissimilar.

Example 2.3

Determine the flux density and electric field strength at a distance of 0.5 m from an isolated point charge of +10 µC. If an identical charge is placed at this point, determine the force it experiences. Assume that the charge is in air.

Solution

Let us place the point charge at the centre of a Gaussian sphere of radius 0.5 m. Now, from Gauss' law the total flux through the sphere is equal to the enclosed charge, i.e.,

$$\psi = 10\,\mu C$$

The area of the Gaussian sphere is $4\pi r^2$, and so the flux density at 0.5 m is

$$D = \frac{10\times10^{-6}}{4\pi 0.5^2}r$$

$$= 3.2\times10^{-6}r\ \mathrm{C\,m^{-2}}$$

The strength of the electric field at this radius is

$$E = \frac{D}{\varepsilon_0}r$$

$$= \frac{3.2\times10^{-6}}{8.854\times10^{-12}}r$$

$$= 3.6\times10^{5}r\ \mathrm{N\,C^{-1}}$$

Now, if we introduce a 10 µC point charge at this distance, the charge will experience a repulsive force of

$$F = qE$$

$$= 10\times10^{-6}\times3.6\times10^{6}r$$

$$= 3.6r\ \mathrm{N}$$

As a matter of interest, if we halve the distance, we get

$$D = \frac{10\times10^{-6}}{4\pi 0.25^2}r$$

$$= 12.7\times10^{-6}r\ \mathrm{C\,m^{-2}}$$

$$E = \frac{12.7\times10^{-6}}{8.854\times10^{-12}}r$$

$$= 1.44\times10^{6}r\ \mathrm{N\,C^{-1}}$$

and

$$F = 1.44 \times 10^6 \times 10 \times 10^{-6} r$$

$$= 1.44r \text{ N}$$

This example has shown that, in spite of the small values of charge, and the large distance between them, the electrostatic force can be quite high.

We can now write three, equivalent, forms of Coulomb's law

$$F = \frac{q_1 q_2}{4\pi \varepsilon r^2} r \text{ N}$$

$$F = \frac{D}{\varepsilon} q_2 \text{ N}$$

and

$$F = q_2 E \text{ N}$$

where electric flux density is

$$D = \frac{q_1}{4\pi r^2} r \text{ C m}^{-2}$$

and electric field strength is

$$E = \frac{D}{\varepsilon} = \frac{q_1}{4\pi \varepsilon r^2} r \text{ N C}^{-1}$$

Let us now turn our attention to electric potential, a term that is usually associated with circuit theory.

2.4   ELECTRIC POTENTIAL

We often come across the term potential when applied to the potential energy of a body or the potential difference between two points in a circuit. In the former case, the potential energy of a body is related to its height above a certain reference level. Thus, a body gains potential energy when we raise it to a higher level. This gain in energy is equal to the work done against an attractive force, gravity in this example. Figure 2.6a shows this situation.

As Figure 2.6a shows, the body is placed in an attractive, gravitational force field. So, if we raise the body through a certain distance, we have to do work against the gravitational field. The difference in potential energy between positions 1 and 2 is equal to the work done in moving the body from 1 to 2, a distance of l metres. This work done is given by

$$F \times l = m \times 9.81 \times l$$

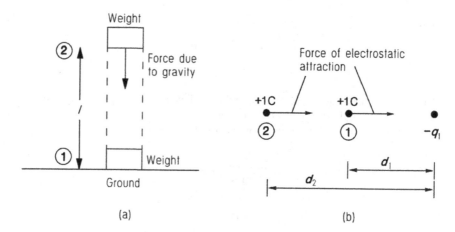

FIGURE 2.6 (a) Potential energy in a gravitational field and (b) potential energy in an electrostatic field.

where m is the mass of the body (kg) and 9.81 is the acceleration due to gravity (m s^{-2}). (Although the effects of gravity vary according to the inverse square law, the difference in gravitational force between positions 1 and 2 is small. This is because the Earth is so large. Thus, we can take the gravitational field to be linear in form, and so this equation holds true.)

In an electrostatic field, we have an electrostatic force field instead of a gravitational force field. However, the idea of potential energy is the same. Let us consider the situation in Figure 2.6b. We have a positive test charge of 1 C at a distance d_1 from the fixed negative charge, $-q_1$. This test charge will experience an attractive force whose magnitude we can find from Coulomb's law. Now, if we move the test charge from position 1 to position 2, we have to do work against the field. If the distance between positions 1 and 2 is reasonably large, the strength of the force field decreases as we move away from the fixed charge. Thus, we say that we have a non-linear field.

As the field decreases when we move away from the fixed charge, let us move the test charge a very small distance, dr. The electric field strength will hardly alter as we move along this small distance. So, the work done against the field in moving the test charge a small distance dr will be given by

$$\text{work done} = \text{force} \times \text{distance}$$

$$= -F \times dr$$

$$= -1 \times E \times dr \qquad (2.8)$$

(The presence of the negative sign is due to the fact that we are moving away from the charge, whereas the electrostatic force acts towards the charge, i.e., in the opposite direction.)

We can move from position 1 to position 2 in very tiny steps so that the E field hardly varies with each step. With each step we take, we will do a small amount of work against the field. To find the total amount of work done, and hence the potential difference, we can integrate Equation (2.8) with respect to r, with d_1 and d_2 as the limits. Thus,

$$\text{total work done} = -\int_{d_1}^{d_2} E \, dr$$

$$= -\int_{d_1}^{d_2} \frac{-q_1}{4\pi\varepsilon r^2} \, dr$$

$$= +\frac{q_1}{4\pi\varepsilon} \int_{d_1}^{d_2} \frac{1}{r^2} \, dr$$

$$= +\frac{q_1}{4\pi\varepsilon} \left\{ \frac{1}{d_1} - \frac{1}{d_2} \right\}$$

In electrical engineering, we use the symbol V for potential. Thus, the potential difference between positions 1 and 2 is

$$V_{12} = +\frac{q_1}{4\pi\varepsilon} \left\{ \frac{1}{d_1} - \frac{1}{d_2} \right\} \tag{2.9}$$

(Although the units of work are joules, we tend to use the volt (named after Count Alessandro Volta, 1745–1827, the Italian physicist who invented the first electric battery in 1800) as the unit of potential.)

Let us take a moment to examine Equation (2.9) more closely. In particular, let us look at the term in the brackets. The major question is whether this term is positive or negative. If d_1 and d_2 are of the same order of magnitude, we might find it difficult to decide. However, if we make d_1 very, very small, and d_2 very, very large, the term in the brackets should be positive. As a check, let us take $d_1 = 0$ and $d_2 = \infty$. Thus,

$$V_{12} = \frac{q_1}{4\pi\varepsilon} \left\{ \frac{1}{0} - \frac{1}{\infty} \right\}$$

$$= \frac{q_1}{4\pi\varepsilon} \{\infty - 0\}$$

$$= \frac{q_1}{4\pi\varepsilon} \infty$$

Although this quantity is clearly very big, it is also very definitely positive. This confirms that we have to do work against the field in moving the test charge away from the negative charge q_1.

Before we consider an example, let us return to Equation (2.9) again. We can recast this equation as

$$V_1 - V_2 = \frac{q_1}{4\pi\varepsilon} \frac{1}{d_1} - \frac{q_1}{4\pi\varepsilon} \frac{1}{d_2}$$

from which we can infer that

$$V_1 = \frac{q_1}{4\pi\varepsilon} \frac{1}{d_1} \tag{2.10}$$

and

$$V_2 = \frac{q_1}{4\pi\varepsilon} \frac{1}{d_2} \tag{2.11}$$

These voltages are the absolute potentials at points 1 and 2, respectively. We can also find these potentials by moving the test charge from infinity to positions 1 and 2. We must take great care with the minus sign when calculating the absolute potential.

Example 2.4

Determine the absolute potential at a distance of 0.2 m from an isolated point charge of 10 μC. Hence, determine the potential difference between this point and another at 10 m from the charge.

Solution

The absolute potential is defined as the work done against the field in moving a positive 1 C test charge from initially to a point in the field. So, the small amount of work done, dV, in moving distance dr is

$$dV = -\text{force} \times dr$$

$$= -1 \times E \times dr$$

$$= -E \ dr$$

(Note that this gives

$$E = -\frac{dV}{dr}$$

and so E can have alternative units of V m⁻¹. We will return to this very important point later.)

Thus, the total work done in moving the charge from infinity to 0.2 m from the fixed charge, the potential, is

$$\int_0^V dV = \int_\infty^{0.2} \frac{10 \times 10^{-6}}{4\pi\varepsilon_0 r^2} dr$$

Therefore,

$$V = -\frac{10 \times 10^{-6}}{4\pi\varepsilon_0} \left| -\frac{1}{r} \right|_\infty^{0.2}$$

$$= -\frac{10 \times 10^{-6}}{4\pi\varepsilon_0} \frac{1}{0.2}$$

$$= 4.5 \times 10^5 \text{ V at } 0.2 \text{ m}$$

By following a similar procedure, the potential at 10 m from the charge is

$$V = 9 \times 10^3 \text{ V}$$

Thus, the potential difference between 0.2 and 10 m is

$$V_{12} = 4.41 \times 10^5 \text{ V}$$

Before we continue, it is worth stressing again that the units of the **E** field are also V m^{-1}. We will use these units when we come to consider capacitors in Section 2.9.

2.5 EQUIPOTENTIAL LINES

Let us consider the three paths A, B and C shown in Figure 2.7a. All of these paths link points 1 and 2, but only path A does so directly. Now, let us take the circular lines in Figure 2.7a as the contours on a hill. In moving from position 1 to position 2 by way of path A, we clearly do work against gravity. The work done is equal to the gain in potential energy which, in turn, is equal to the gravitational force times the change in vertical height. (This is shown in Figure 2.7b.)

Now let us take path B. We initially walk left from position 1, around the contour line, to a point directly below position 2. As we have moved around a contour line, we have not gained any height, and so the potential energy remains the same, i.e., we have not done any work against gravity. We now have to walk uphill to position 2. In doing so we do work against gravity equal to the gain in potential energy. This gain in potential energy is clearly the same as with path A. (Although we have to do more physical work in travelling along path B, the change in potential energy is the same.) If we use path C, the same argument holds true. So, we can say that the work done against gravity is independent of the path we take.

Let us now turn our attention to the electrostatic field in Figure 2.8. As with the contour map, we have three different paths. As we have just seen, we do no work against the field when we move in a circular direction. We only do work when we move in a radial direction. Thus, the potential difference between points 1 and 2 is

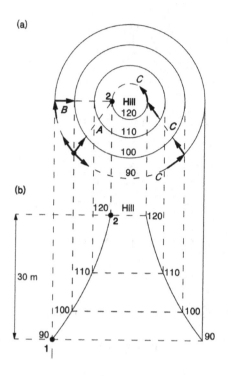

FIGURE 2.7   (a) Contour map for a circular hill and (b) side view of hill.

independent of the exact path we take. This implies that we do no work against the field when we move around the plot in a circular direction. Thus, the circular 'contours' in Figure 2.8 are lines of equal potential or equipotential lines.

We should be careful when using the term equipotential lines. This is because we are considering a point charge, and so the equipotential surfaces are actually spheres with the charge at their centre. As we are not yet able to draw in a three-dimensional holographic world, we have to make do with two-dimensional diagrams drawn on pieces of paper!

Example 2.5

An isolated point charge, of 20 pC, is situated in air. Plot the 1, 2 and 3 V equipotentials.

Solution

To plot the equipotentials, we need to find the radius of the 1, 2 and 3 V potential spheres. Now, the absolute potential of a point in an **E** field is

$$V = \frac{q}{4\pi\varepsilon_0} \frac{1}{r}$$

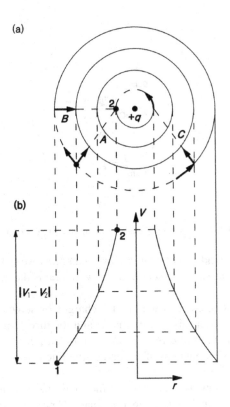

FIGURE 2.8 (a) 'Contour' map for a positive point charge and (b) variation in potential as a function of distance from charge.

and so,

$$r = \frac{q}{4\pi\varepsilon_0 v}$$

Thus,

$$r_{1 v} = 18 \text{ cm}$$

$$r_{2 v} = 9 \text{ cm and, } r_{3 v} = 6 \text{ cm}$$

These lines are plotted in Figure 2.9.

2.6 LINE CHARGES

So far we have only considered point charges. This is very useful when introducing the fundamental laws we have considered. However, we rarely meet point charges in reality. Instead, we come across lines of charge, charged surfaces and charged objects. Thus, we have to deal with charge distributions in one, two and three dimensions.

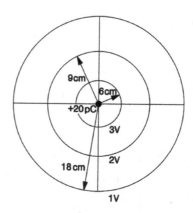

FIGURE 2.9   1, 2 and 3 V equipotentials surrounding a +20 pC point charge.

This is where things tend to get a little complicated, as we have to think in three dimensions. In such cases, it is essential to draw diagrams that help us visualize the situation.

Let us consider a long piece of wire that is charged by some means. Electric flux will radiate outwards from this line of charge and the direction of this flux will be away from the line in a radial direction. If we only consider the central part of the wire, we can ignore what happens at the end of the line and so the flux distribution is as shown in Figure 2.10.

If we apply Gauss' law, we can say that the total flux emanating from the wire is equal to the charge enclosed by an imaginary Gaussian surface. In this case, the Gaussian surface will be an open-ended tube with the wire placed along the central axis of the tube (Figure 2.10). To find the flux density, and hence the electric field strength, we can use Gauss' law or we can use a more rigorous mathematical approach. Both techniques are presented here.

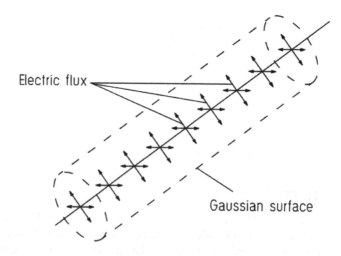

FIGURE 2.10   Radiation of electric flux from a line of charge.

2.6.1 GAUSS' LAW APPROACH

Let us consider the line charge and Gaussian surface as shown in Figure 2.10. The charge is distributed along the length of the wire, and so let us introduce a line charge density given by the total charge, Q, divided by the length of the line, L, i.e.,

$$\rho_1 = \frac{Q}{L} \tag{2.12}$$

If we consider a unit length of wire (1 m), we get a total flux of

$$\psi = \rho_1 \times l \; \text{C} \tag{2.13}$$

Now, the flux density is the flux divided by the surface area of the Gaussian surface. As the Gaussian surface is a tube, the surface area is the circumference of the tube times the length, i.e.,

$$\text{area} = 2\pi r \times l$$

Thus, the density is

$$D = \frac{\rho_1}{2\pi r} r \tag{2.14}$$

and the electric field strength is

$$E = \frac{\rho_1}{2\pi \varepsilon r} r \tag{2.15}$$

The equipotential surfaces will be coaxial tubes that have the wire along the centre line of the tubes (Figure 2.11a). So, if we move in a direction parallel to the wire, we do no work against the field indicating that we can ignore travel along the wire. Thus, we can draw a two-dimensional plot as shown in Figure 2.11b.

If we move a 1 C test charge a small distance in the E field, the small amount of work done is

$$dV = -E \, dr$$

Thus, the total amount of work done against the E field in moving the test charge from infinity to a point in the field is

$$\int_0^V dV = -\int_\infty^R \frac{\rho_1}{2\pi \varepsilon r} \, dr$$

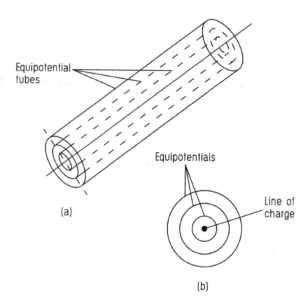

FIGURE 2.11 (a) Equipotential tubes surrounding a line charge and (b) two-dimensional plot of equipotentials.

Unfortunately, solution of this equation results in an infinite potential! (Readers might like to try this for themselves.) The way round this is to set the potential to zero at a distance r, where r tends towards infinity. Thus,

$$\int_0^V dV = -\int_r^R \frac{\rho_1}{2\pi\varepsilon r}\,dr \tag{2.16}$$

as $r \to \infty$. Fortunately, we are more usually concerned with potential differences, and so the dummy variable r, in Equation (2.16), cancels out as shown in the following example.

Example 2.6

A 10-m long wire has a charge of 20 μC along it. Determine the flux density and the electric field strength at a radial distance of 0.5 m from the wire. In addition, find the potential difference between points at 0.5 and 1.5 m from the wire.

Solution

The line is 10 m long with a charge of 20 μC. Thus, the charge density is

$$\rho_1 = \frac{20 \times 10^{-6}}{10}$$

$$= 2\ \mu C\ m^{-1}$$

Let us now consider a section of the wire length l metre. The charge on this length is

$$q = 2 \times 10^{-6} \times l$$

$$= 2 \times l \ \mu C$$

We want to find the flux density at a radius of 0.5 m from the wire. The surface area of the Gaussian surface will therefore be

$$\text{area} = 2\pi \times 0.5 \times l \ \text{m}^2$$

and so the flux density is

$$D = \frac{2 \times 10^{-6} \times l}{2\pi \times 0.5 \times l} r$$

$$= 6.4 \times 10^{-7} r \ \text{C m}^{-2}$$

(It is worth noting that the length of the wire cancels out, and so we could have taken any length we liked.) The electric field strength is given by

$$E = \frac{D}{\varepsilon_0}$$

$$= 7.2 \times 10^4 r \ \text{V m}^{-1}$$

$$= 72r \ \text{kV m}^{-1}$$

As regards the potential difference, we can use Equation (2.16) to give

$$V_{0.5} = \frac{2 \times 10^{-6}}{2\pi\varepsilon} (\ln r - \ln 0.5)$$

and

$$V_{1.5} = \frac{2 \times 10^{-6}}{2\pi\varepsilon} (\ln r - \ln 1.5)$$

with r tending to infinity. Therefore,

$$V_{0.5} - V_{1.5} = \frac{2 \times 10^{-6}}{2\pi\varepsilon} ((\ln r - \ln 0.5) - (\ln r - \ln 1.5))$$

$$= \frac{2 \times 10^{-6}}{2\pi\varepsilon} (\ln 1.5 - \ln 0.5)$$

$$= \frac{2 \times 10^{-6}}{2\pi\varepsilon} \ln 3$$

$$= 40 \ \text{kV}$$

2.6.2 MATHEMATICAL APPROACH

In the previous section, we used Gauss' law to examine the field around an infinitely long charged line. Here, we will split the line into infinitesimally small sections, so that we can use Coulomb's law as applied to point charges.

Let us consider a small section of the line as shown in Figure 2.12. The charge on this section is $\rho_1 \times dz$ and so the flux density at point P is

$$D = \frac{\rho_1 \times dz}{4\pi r^2} \qquad (2.17)$$

We should note that this is a vector quantity acting at an angle to the axis of the line. By taking the origin as shown in Figure 2.12, the line stretches from $-\infty$ to $+\infty$. Thus, we can see that the line is symmetrical about the origin. Now, D can be split into a radial component, D_r, and a vertical component, D_z, given by

$$D_r = \frac{\rho_1 \times dz}{4\pi r^2} \sin\theta r \qquad (2.18)$$

and

$$D_z = \frac{\rho_1 \times dz}{4\pi r^2} \cos\theta z \qquad (2.19)$$

These flux densities are due to an incremental section of the line. So, to find the total flux density, we can integrate Equations (2.18) and (2.19) with respect to z between the limits of $-\infty$ and $+\infty$. Unfortunately, as we move up and down the line, r and θ vary as well. If we change variables to integrate with respect to θ, we need to express z and r in terms of θ. So, to return to Equation (2.18), we have

$$D_r = \frac{\rho_1 \times dz}{4\pi r^2} \sin\theta r$$

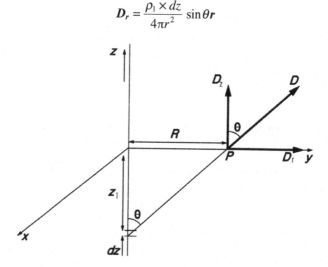

FIGURE 2.12   Field at a point P due to a small section of line.

Now,

$$\tan \theta = \frac{R}{z_1}$$

which, after differentiation, becomes

$$\frac{d\theta}{\cos^2 \theta} = \frac{R}{z_1} dz$$

and so,

$$dz = \frac{z_1^2 \, d\theta}{R \cos^2 \theta}$$

Thus, Equation (2.18) becomes

$$\boldsymbol{D}_r = \frac{\rho_1}{4\pi} \times \frac{\sin \theta}{r^2} \times \frac{-z_1^2}{R \cos^2 \theta} d\theta \ \boldsymbol{r}$$

Now, $\cos \theta = z_1/r$ and so we can write

$$\boldsymbol{D}_r = \frac{\rho_1}{4\pi} \times \sin \theta \times \frac{\cos^2 \theta}{R \cos^2 \theta} d\theta \ \boldsymbol{r}$$

$$= -\frac{\rho_1}{4\pi R} \sin \theta d\theta \ \boldsymbol{r} \qquad (2.20)$$

To find the total radial flux density, we need to integrate this equation with respect to θ. As this is an infinite line, the limits of θ are 0 and π. Thus,

$$\boldsymbol{D}_r = -\frac{\rho_1}{4\pi R} \int\limits_0^\pi \sin \theta d\theta \ \boldsymbol{r}$$

$$= -\frac{\rho_1}{4\pi R} \left| -\cos \theta \ d\theta \ \right|_0^\pi \boldsymbol{r}$$

$$= -\frac{\rho_1}{4\pi R} (-2) \boldsymbol{r}$$

$$= -\frac{\rho_1}{2\pi R} \boldsymbol{r} \qquad (2.21)$$

This is exactly the same as the radial field given by Equation (2.14). However, what about the flux density in the z-direction? Well, the line is symmetrical about $z = 0$. Thus, there will be an identical component of D_z, acting in the opposite direction, due to an incremental section at $z = -z_1$. Hence, we can say that the axial component of D will be zero. (Readers can confirm this for themselves by integrating Equation (2.19) with respect to θ.) As regards the E field, and the potential, we can follow an identical procedure to that used in the previous section.

So, this method has resulted in exactly the same result as that obtained using Gauss' law. Although this derivation has involved us in a considerable amount of work, it has introduced us to the question of symmetry, and the resultant simplifications it can bring.

2.7 SURFACE CHARGES

The last section concentrated on line charges that we find when we have a charged wire. We also come across surface charges, such as those on capacitors and electrostatic precipitators. In such cases, we can again use a mathematical approach, or we can apply Gauss' law. In common with the previous section, we will use both approaches.

2.7.1 GAUSS' LAW APPROACH

Let us consider the circular charged plate shown in Figure 2.13. This plate has a certain charge spread over its surface. To simplify the analysis, let us assume that the charge distribution is uniform and that there are no edge effects. Now let us consider a small area of the plate. This area will contain a certain amount of charge, dQ, given by

$$dQ = \rho_s \, ds \qquad\qquad (2.22)$$

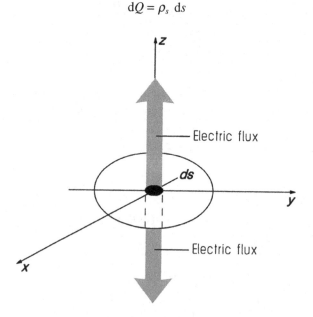

FIGURE 2.13   Radiation of electric flux from a charged surface.

where ρ_s is the surface charge density in C m^{-2}, and ds is the area of the section. Flux emanating from this area will flow upwards and downwards to occupy a cylinder. (There will only be a vertical component of flux because any horizontal flux will cancel out due to symmetry. This is a similar situation to that which we met when we examined line charges.) By applying Gauss' law, we see that the total flux out of the cylinder, in both directions, must equal the enclosed charge. So, half the flux flows upwards and half flows downwards. Thus, the flux density at any height above the disc is

$$D_z = \frac{dQ}{2ds}z$$

$$= \frac{\rho_s ds}{2ds}z$$

$$= \frac{\rho_s}{2}z \qquad (2.23)$$

and the electric field strength is

$$E_z = \frac{\rho_s}{2\varepsilon_0\varepsilon_r}z \qquad (2.24)$$

The important thing to note here is that the E field is independent of the distance from the disc. This is a consequence of having the flux flow in a cylindrical tube.

As regards the potential, we have equipotential surfaces parallel to the disc. These surfaces will have the same area as the disc. If we take zero potential at infinity, the absolute potential at a distance z from the disc will be

$$\int_0^V dV = \int_\infty^z \frac{\rho_s}{2\varepsilon_0\varepsilon_r}dz$$

Therefore,

$$V = \frac{\rho_s}{2\varepsilon_0\varepsilon_r}z \qquad (2.25)$$

This is an example of a linear field because the E field is constant regardless of the distance from the disc. We should remember, however, that this result has only appeared because we have a uniform charge density.

Example 2.7

Determine the flux density, and hence the electric field strength, produced at a distance of 1 m from the centre of a 1 m^2 square of insulating material that has a total charge of 10 pC evenly distributed over it. Assume that the square is in air.

Solution

Let us place the square at the centre of a set of Cartesian axes as shown in Figure 2.14. The square lies in the xy-plane, and so the electric flux will act equally along the positive and negative z-axis. As we have just seen, the electric flux density is independent of the exact shape of the material and is given by Equation (2.23) as

$$D_z = \frac{\rho_1}{2} z$$

Thus,

$$D_z = \frac{10 \times 10^{-12}}{2} z$$

$$= 5z \ \text{pCm}^{-2}$$

As regards the E field, we can use $D = \epsilon E$ to give

$$E_z = \frac{5 \times 10^{-12}}{8.854 \times 10^{-12}} z$$

$$= 0.56 \ z \ \text{V m}^{-1}$$

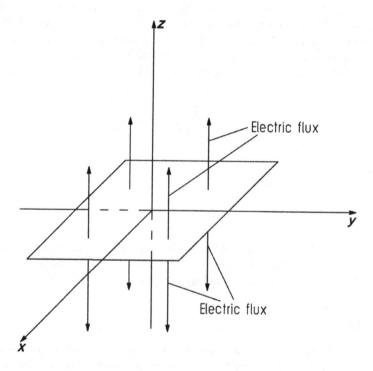

FIGURE 2.14 Radiation of electric flux from a charged square.

2.7.2 MATHEMATICAL APPROACH

As we have previously seen, we can only apply Coulomb's law, and hence use our usual expressions for D and E, when considering point charges. However, we have a surface charge, and so how can we analyze this situation? The solution is to consider a small section of the disc, calculate the flux density due to the charge on this small section and integrate the result over the area of the disc. (Problem 1.2 gives more information about the integration method used in the following derivation.)

As Figure 2.15 shows, let us consider a small section of a disc. The area of this small section is

$$ds = rd\phi\ dr \qquad\qquad (2.26)$$

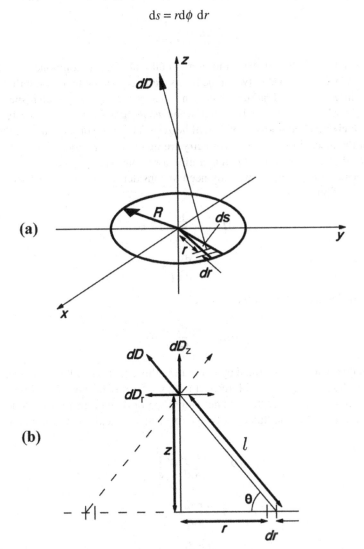

FIGURE 2.15   (a) Field at point P due to a small section of a charged disc and (b) side view.

if $d\phi$ is expressed in radians. (This is a direct consequence of expressing $d\phi$ in radians. The circumference of the disc is $2\pi r$, and this encloses an angle of 360°, or 2π radians. So, the length of an arc that subtends an angle of 180°, or π radians, is πr. Thus, the length of an arc is equal to the product of the angle (in radians) and the radius of the arc.)

If we have a charge spread over the surface of this disc, the charge on this small section is

$$dQ = p_s ds \tag{2.27}$$

which we can take to be a point charge if ds is very small. This charge will produce a small component of the flux density acting in the direction shown in Figure 2.15a. So,

$$dD = \frac{p_s ds}{4\pi l^2} \tag{2.28}$$

We can resolve this flux density into horizontal and vertical components, and then integrate with respect to Φ between the limits 0 and 2π. This integration will describe a ring of thickness dr and radius r. It is then a matter of integrating with respect to r to map out the whole of the disc. However, when we integrate with respect to Φ, we find that the **horizontal** component of D will be zero. This is a consequence of the symmetry of the disc, similar to the symmetry we met in the previous section. (Readers can check this by performing the integration for themselves.)

So, because of symmetry, we only need to consider the vertical component of the flux density. Thus,

$$dD_z = \frac{p_s ds}{4\pi l^2} \sin\theta \; z$$

$$= \frac{p_s ds}{4\pi l^2} \frac{z}{l} z$$

Now, from Equation (2.26), we can write

$$dD_z = \frac{p_s r d\phi \; dr}{4\pi l^2} \frac{z}{l} z \tag{2.29}$$

The total flux density is obtained by integrating this equation with respect to Φ and r. We can perform integration with respect to Φ very easily because l does not vary as we move in a circular direction. (The integration of ds with respect to Φ describes a ring of radius r.) So, the flux density due to a ring of thickness dr and radius r is

$$dD_z = \frac{p_s r \; dr}{4\pi l^3} \frac{z}{} \int_0^{2\pi} d\phi \; z$$

$$= \frac{p_s r \; dr}{4\pi l^3} \frac{z}{} 2\pi z$$

$$= \frac{p_s r \; dr}{2 l^3} z \tag{2.30}$$

We now need to integrate Equation (2.30) with respect to radius to find the total flux density at the point P. Unfortunately, as we perform the integration, the length l varies as the radius goes from 0 to R. Thus, we need to express l in terms of r prior to integrating with respect to r. So,

$$D_z = \frac{\rho_s z}{2} \int_0^R \frac{r\ dr}{l^3}\ z$$

$$= \frac{\rho_s z}{2} \int_0^R \frac{r\ dr}{\left(r^2 + z^2\right)^{3/2}}\ z$$

$$= \frac{\rho_s z}{2} \left. \frac{-1}{\left(r^2 + z^2\right)^{1/2}} \right|_0^R\ z$$

i.e.,

$$D_z = \frac{\rho_s z}{2} \left[\frac{1}{z} - \frac{1}{\left(R^2 + z^2\right)^{1/2}} \right] z \tag{2.31}$$

We can perform a very simple check on this equation by letting $z \to \infty$ (i.e. find the flux density at infinity). Under these circumstances, we would expect the disc to approximate to a point charge, and so the flux density should approximate to Equation (2.4). So, by using the binomial expansion of the term in brackets, we get

$$D_z \approx \frac{\rho_s z}{2} \left(\frac{1}{z} - \frac{1}{z} + \frac{1}{2z} \frac{R^2}{z^2} \right) z$$

$$= \frac{\rho_s z}{2} \frac{1}{2z} \frac{R^2}{z^2}\ z$$

$$= \rho_s \frac{R^2}{4z^2}\ z$$

$$= \frac{Q}{\pi R^2} \frac{R^2}{4z^2}\ z$$

$$= \frac{Q}{4\pi z^2}\ z$$

which is the equation for the flux density resulting from a point charge.

We can perform another simple check by letting $R \rightarrow \infty$. If we do this, we find from Equation (2.31), that

$$D_t \rightarrow \frac{\rho_s}{2} z$$

which is the same as that found using Gauss's law, Equation (2.23).

So, we have successfully shown that the flux density from a surface charge distribution reduces to that of a point charge, if the distance from the surface is large enough. We have also shown that the mathematical approach gives the same result as that obtained using Gauss' law.

2.8 VOLUME CHARGES

The previous two sections have introduced us to line and surface charge densities. However, we live in a three-dimensional world (four if you count time) and so we will often meet charged volumes. When considering a volume charge density, we can make good use of Gauss' law to replace the volume charge by a point charge at the centre of the volume.

As an example, let us consider a sphere with a charge evenly distributed throughout its volume, as shown in Figure 2.16a. To analyze this situation, we could consider a small section of the sphere, and perform an integration to map out the whole of the volume. However, this will involve us in a considerable amount of work. An alternative is to apply Gauss' law and replace the volume charge density by a point charge at the centre of the volume.

So, if the sphere has a total charge of coulomb distributed throughout the volume, we can replace the sphere by a point charge, placed at the centre placed the centre of the sphere, of magnitude Q. This is shown in Figure 2.16b. It is then a simple to find the flux density, etc., at any distance from the surface of the sphere.

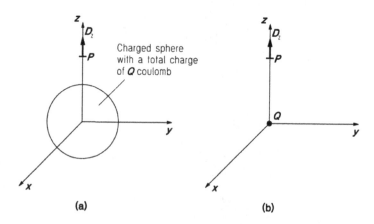

FIGURE 2.16 (a) Field at point P due to a charged sphere and (b) simplification due to application of Gauss' law.

Example 2.8

A solid sphere, of radius 0.25 m, has a charge of 10 pC evenly distributed through-out its volume. By using Gauss' law, determine the flux density at point 1 m from the centre of the sphere.

Solution

We have to apply Gauss' law to the sphere. Now, the total charge contained throughout the sphere is 10 pC. So, we can replace the sphere by a point charge of 10 pC placed at the centre of the sphere. We now need to determine the flux density at a distance of 1 m from this charge. Thus,

$$D_r = \frac{10 \times 10^{-12}}{4\pi l^2} r$$

$$= 0.8r \text{ pC m}^{-2}$$

Although we could have considered the flux due to a small incremental volume of the sphere, and then integrated throughout the volume of the sphere, the math-ematics would be very complicated indeed. (Interested readers can show this for themselves, but it is not recommended!)

2.9   CAPACITORS

Most of us are familiar with capacitors as circuit elements and, as such, we seldom need to examine their structure. What is not often realized is that a capacitor is formed whenever we have two conductors close to each other. Such a situation fre-quently occurs in electrical engineering, but the effect does not make itself felt until we reach high frequencies.

In this section, we will consider parallel plate capacitors, coaxial cable, twin feeder and microstrip line. All of these capacitors are commonly found in electrical engineering. We begin our study by examining the parallel plate capacitor.

2.9.1   PARALLEL PLATE CAPACITORS

Capacitors come in a variety of shapes and sizes: from big electrolytic capacitors for smoothing the output of power supplies to small value disc ceramic capacitors for use in high-frequency circuits. All types of capacitor are based on the simple structure shown in Figure 2.17.

For convenience, we will consider a circular plate capacitor with R being the radius of each plate. When the bottom plate of the capacitor has a charge on it, this charge induces an equal and opposite charge in the top plate. Thus, if the bottom electrode has a charge of $+Q$ on it, the charge on the upper plate is $-Q$.

Now, as we saw in Section 2.7, if the charge is evenly distributed over the plate, the flux is equally divided into upward and downward flux. So, flux flows upwards through the dielectric from the positively charged lower plate. As we have an upper

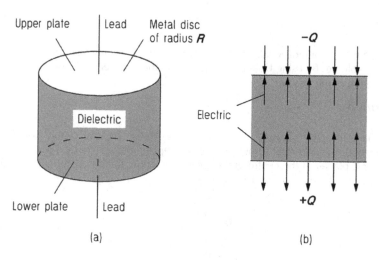

FIGURE 2.17 (a) Basic structure of a parallel plate capacitor and (b) flux distribution in a parallel plate capacitor.

plate with a charge of $-Q$ on it, flux in the upper half of the capacitor will also flow upwards. Thus, the total flux in the capacitor flows in the positive z-direction as shown in Figure 2.17b.

As we have two equal sources of flux acting in the capacitor, the total flux density in the capacitor is

$$D_z = \left(\frac{\rho_s}{z} + \frac{\rho_s}{2} \right) z$$

$$= \frac{Q}{\pi R^2} z$$

(2.32)

and so the electric field strength is

$$E_z = \frac{Q}{\varepsilon_0 \varepsilon_r \pi R^2} z$$

(2.33)

Thus, the potential difference between the top and bottom plate is

$$\int_{V_1}^{V_2} dV = \frac{-Q}{\varepsilon_0 \varepsilon_r \pi R^2} \int_{-d/2}^{d/2} - dz$$

and so,

$$V_2 - V_1 = \frac{Q}{\varepsilon_0 \varepsilon_r \pi R^2} \left(\frac{d}{2} + \frac{d}{2} \right)$$

$$= \frac{Q}{\varepsilon_0 \varepsilon_r \pi} \frac{d}{R^2}$$

or, more generally,

$$V = \frac{Q}{\varepsilon_0 \varepsilon_r \pi R^2} d$$

We can rearrange this equation to give

$$Q = \frac{\varepsilon_0 \varepsilon_r \pi R^2}{d} V$$

which shows that the stored charge is directly proportional to the voltage across the capacitor plates. The constant of proportionality is the capacitance given by

$$C = \frac{\varepsilon_0 \varepsilon_r \times \text{area}}{d} \qquad (2.34)$$

and so we can also write

$$Q = CV \qquad (2.35)$$

Both equations should be familiar to most readers.

2.9.2 COAXIAL CABLE

Coaxial cable is very widely used in everyday life: a familiar example is the lead that connects the aerial to the television set. Figure 2.18 shows the basic structure of this type of cable.

Under normal circumstances, the inner conductor of the cable carries the signal, or voltage, whereas the outer conductor is usually earthed. The advantage of this structure is that any external interference has to pass through an earthed conductor before it reaches the signal. In effect, the outer conductor shields the signal from any external interference.

Now, if the inner conductor is at a certain potential above the earthed shield, we will have a capacitor. To find the capacitance, we need an equation that links the potential difference between inner and outer conductors to the charge on the inner conductor.

As we are dealing with a length of cable, let us assume that the inner conductor has a charge of ρ_1 coulomb per unit length. This charge will produce flux in a radial direction similar to that which we met in Section 2.6. The Gaussian surface in this instance is a tube of radius r, thickness dr and length l.

So, the flux flowing through this surface is

$$\psi = \rho_1 \times l$$

As this flux flows in a radial direction, the flux density through the Gaussian surface is

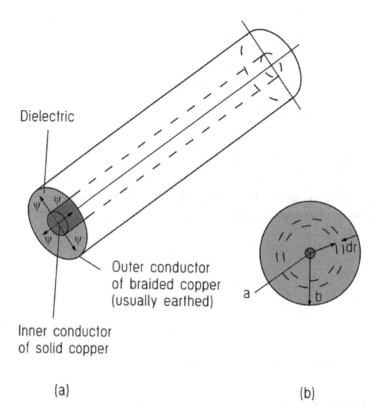

Dielectric

Outer conductor
of braided copper
(usually earthed)

Inner conductor
of solid copper

(a)   (b)

FIGURE 2.18   (a) Basic structure of coaxial cable and (b) end view of coaxial cable.

$$D = \frac{\rho_1 l}{2\pi r l}$$

or

$$D_r = \frac{\rho_1}{2\pi r} r$$

and so the electric field strength at this radius is

$$E_r = \frac{\rho_1}{2\pi\varepsilon_0\varepsilon_r r} r$$

Now, the thickness of the Gaussian surface is dr, and so the potential difference across the surface is

$$dV = -E_r dr$$

$$= -\frac{\rho_1}{2\pi\varepsilon_0\varepsilon_r}\frac{1}{r}dr$$

The potential difference between the inner and outer conductors is therefore given by

$$\int_0^V dV = - \frac{\rho_1}{2\pi\varepsilon_0\varepsilon_r} \int_b^a \frac{1}{r}\,dr$$

and so,

$$V = - \frac{\rho_1}{2\pi\varepsilon_0\varepsilon_r} \left| \ln r \right|_b^a$$

$$= \frac{\rho_1}{2\pi\varepsilon_0\varepsilon_r} \ln\left(\frac{b}{a}\right)$$

$$= \frac{Q}{2\pi\varepsilon_0\varepsilon_r l} \ln\left(\frac{b}{a}\right)$$

Thus, the capacitance is

$$C = \frac{Q}{V}$$

$$= \frac{2\pi\varepsilon_0\varepsilon_r}{\ln\left(\frac{b}{a}\right)} \times \text{length} \qquad (2.36)$$

and so the capacitance per unit length is

$$C' = \frac{2\pi\varepsilon_0\varepsilon_r}{\ln\left(\frac{b}{a}\right)} \; \text{F m}^{-1} \qquad (2.37)$$

We should note that the capacitance per unit length is directly dependent on the length of the cable. So, if we double the length of the cable, the capacitance also doubles.

Example 2.9

A 500 m length of coaxial cable has an inner conductor of radius 2 mm and an outer conductor of radius 1 cm. The relative permittivity of the dielectric separating the inner and outer conductor is 5. Determine the capacitance of the cable. If the inner conductor is at a potential of 1 kV above the outer conductor, determine the maximum value of the E field in the dielectric.

Solution

We want to find the capacitance of the cable. So, we can use Equation (2.36) to give

$$C = \frac{2\pi\varepsilon_0\varepsilon_r}{\ln\left(\dfrac{b}{a}\right)} \times \text{length}$$

$$= \frac{2\pi \times 8.854 \times 10^{-12} \times 5}{\ln\left(\dfrac{10^{-2}}{2} \times 10^{-3}\right)} \times 500$$

and so,

$$C' = 173\,\text{pF m}^{-1}$$

As regards the **E** field in the dielectric, we have seen that

$$E_r = \frac{\rho_1}{2\pi\varepsilon_0\varepsilon_r r}\, r$$

In order to find E_r we need to know the charge per unit length, ρ_1. As $Q = CV$, Equation (2.35), we can write

$$\rho_1 = C'V$$

$$= 173 \times 10^{-12} \times 1 \times 10^3$$

$$= 173\ \text{nC m}^{-1}$$

As the E field is inversely proportional to radius, the maximum field occurs at the surface of the inner conductor. Thus,

$$E_r/\big|_{\text{max}} = \frac{173 \times 10^{-9}}{2\pi \times 8.854 \times 10^{-12} \times 5 \times 2 \times 10^{-3}}$$

$$= 311\ \text{kV m}^{-1}$$

This is quite a considerable field strength, and one that may cause the dielectric to break down. Chapter 6 deals with this in greater detail.

2.9.3 TWIN FEEDER

In communication, twin feeder is often used as connecting wire between short-wave transmitters and their aerials. We also find twin feeder in power transmission systems and telephone lines. As Figure 2.19 shows, twin feeder generally consists of two parallel wires held apart by some means.

Let us assume that the left-hand conductor has a charge per unit length of p_1 C m^{-1}. This charge will induce an equal and opposite charge on the right-hand conductor.

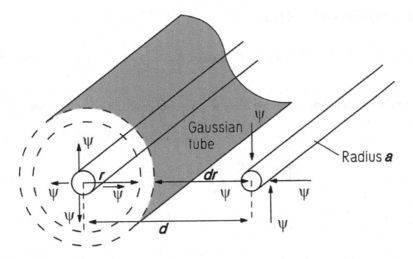

FIGURE 2.19 Basic structure of twin feeder.

The flux emanating from the left-hand conductor does so in a radial fashion, and so we can take a Gaussian tube of length l, similar to that which we used in Section 2.6.

The charge on a length of the left-hand conductor is $p_1 \times l$ coulomb. Thus, the total flux through the Gaussian surface is

$$\psi = \rho_l \times l \qquad (2.38)$$

The area of the Gaussian surface is $2\pi r \times l$ and so the flux density at a radius r is

$$D_r = \frac{\rho_1 \times l}{2\pi r \times l} r$$

$$= \frac{\rho_1}{2\pi r} r \qquad (2.39)$$

Thus, the electric field strength at this radius is

$$E_r = \frac{\rho_1}{2\pi\varepsilon_0 r} r \qquad (2.40)$$

We also have a right-hand conductor with an equal and opposite charge. The radial electric field from this conductor will have the same direction as the field due to the left-hand conductor. The electric field strength due to the right-hand conductor is

$$E_r = \frac{\rho_1}{2\pi\varepsilon_0(d-r)} r \qquad (2.41)$$

and so the total electric field strength is

$$E_r = \frac{\rho_1}{2\pi\varepsilon_0 r} + \frac{\rho_1}{2\pi\varepsilon_0 (d-r)} r \tag{2.42}$$

To find the potential between two lines, we must integrate this equation with respect to radius. So,

$$\int_{V_b}^{V_a} dV = \frac{-\rho_1}{2\pi\varepsilon_0} \int_{d-a}^{a} \left(\frac{1}{r} + \frac{1}{(d-r)} \right) dr$$

Therefore,

$$V_a - V_b = \frac{-\rho_1}{2\pi\varepsilon_0} \left| \ln r - \ln(d-r) \right|_{d-a}^{a}$$

$$= \frac{-\rho_1}{2\pi\varepsilon_0} \left(\ln a - \ln(d-a) + \ln a - \ln(d-a) \right)$$

$$= \frac{-\rho_1}{\pi\varepsilon_0} \left(\ln a - \ln(d-a) \right)$$

$$= \frac{\rho_1}{2\pi\varepsilon_0} \left(\ln(d-a) + \ln a \right)$$

$$= \frac{\rho_1}{\pi\varepsilon_0} \ln\left(\frac{d-a}{a} \right)$$

$$= \frac{Q}{\pi\varepsilon_0} \ln\left(\frac{d-a}{a} \right) \times \text{length} \tag{2.43}$$

So, the capacitance of the arrangement is

$$C = \frac{\pi\varepsilon_0}{\ln|(d-a)/a|} \times \text{length F}$$

or

$$C' = \frac{\pi\varepsilon_0}{\ln|(d-a)/a|} \text{F m}^{-1} \tag{2.44}$$

If the separation of the conductors is significantly greater than the diameter of the conductors, Equation (2.44) reduces to

$$C' = \frac{\pi \varepsilon_0}{\ln\left(\dfrac{d}{a}\right)} \, \text{F m}^{-1} \tag{2.45}$$

Example 2.10

A 200 m length of feeder consists of 2-mm radius conductors separated by a distance of 20 cm. Determine the capacitance of the arrangement.

Solution

As the distance between the conductors is very much greater than the radius of the conductors, we can use Equation (2.45) to give

$$C' = \frac{\pi \varepsilon_0}{\ln\left(\dfrac{d}{a}\right)}$$

$$= \frac{\pi \times 8.854 \times 10^{-12}}{\ln\left(20 \times 10^{-2} / 2 \times 10^{-3}\right)}$$

$$= 6 \, \text{pF m}^{-1}$$

$$= 1.2 \text{ nF} \quad \text{for the 200 m length}$$

2.9.4 WIRE OVER GROUND – THE METHOD OF IMAGES

In the last section, we studied twin feeder. However, we often come across conductors placed over a ground-plane. Under these circumstances, we can make use of the method of images to find the capacitance of the arrangement.

Figure 2.20a shows the situation we are considering. As we can see, we have a line above an infinite ground-plane. (A ground-plane is simply a large conducting area that is earthed. The Earth itself is one example. Another example is the use of double-sided printed circuit boards.) To analyze this situation, we can introduce an imaginary conductor on the other side of the ground-plane, as shown in Figure 2.20b.

As the ground-plane is exactly half-way between the two conductors, it is effectively lying along an equipotential line. This means that we can remove the ground-plane, so leaving us with the twin feeder we have just considered. As we can see from Figure 2.20c, we can consider the two-wire situation as being made up of two single-wire/ground-plane arrangements in series. Thus, the capacitance of the single conductor over a ground-plane is simply twice that of the twin feeder. We can now write

$$C' = \frac{2\pi \varepsilon_0}{\ln\left(\dfrac{\dfrac{d}{2} - a}{a}\right)} \, \text{F m}^{-1} \tag{2.46}$$

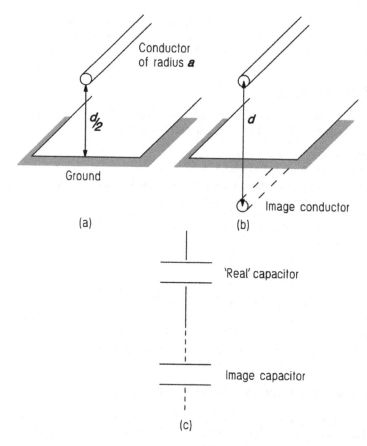

FIGURE 2.20 (a) Wire over ground, (b) image conductor for wire over ground and (c) equivalent circuit of wire and image.

or

$$C' = \frac{2\pi\varepsilon_0}{\ln(d/2a)} \, \text{F m}^{-1} \qquad (2.47)$$

where $d/2$ is the height of the conductor above the ground-plane.

Example 2.11

A high-voltage power line consists of 1-cm radius copper wire placed 25 m above the ground. Determine the capacitance that the line has to the ground.

Solution

The height of the wire above the ground is significantly greater than the radius of the wire, and so we can use Equation (2.47) to give

$$C' = \frac{2\pi\varepsilon_0}{\ln(d/2a)} \, \mathrm{F\,m^{-1}}$$

$$= \frac{2\pi \times 8.854 \times 10^{-12}}{\ln\left(25/2\times10^{-3}\right)}$$

$$= 7.8\,\mathrm{pF\,m^{-1}}$$

2.9.5 Microstrip Line

When constructing electronic circuits, we often use double-sided printed circuit board. When using this type of board, the upper layer generally carries the signal, while the bottom layer is usually earthed. Under these circumstances, the signal line and earth form a capacitor. When operating at low frequencies, the effect is not very pronounced. However, at high frequencies, the capacitance has a greater effect.

Figure 2.21 shows the cross section through a double-sided circuit board. In common with the previous example, we can make good use of the method of images to produce a parallel plate capacitor. Thus, the capacitance of the microstrip will be

$$C = \frac{2\varepsilon_0\varepsilon_r \times \mathrm{area}}{2h}$$

or

$$C' = \frac{\varepsilon_0\varepsilon_r w}{h}\,\mathrm{F\,m^{-1}} \tag{2.48}$$

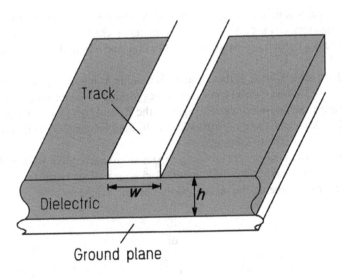

FIGURE 2.21 Cross-section through a double-sided printed circuit board.

If the width of the track is much smaller than the thickness of the board, we can approximate the distribution to that of a cylindrical wire over a ground-plane. This is identical to the situation we met in the last section. Thus, the capacitance per unit length is

$$C' = \frac{2\varepsilon_0 \varepsilon_r}{\ln\left(\dfrac{h}{w}\right)} \, \mathrm{F\,m^{-1}} \tag{2.49}$$

Example 2.12

A 3-mm wide track is etched on one side of some double-sided printed circuit board. The thickness of the board is 2 mm, and the dielectric has a relative permittivity of 5. Determine the capacitance per cm.

Solution

As the track width is of the same order of magnitude as the board thickness, we must use Equation (2.48) to give

$$C' = \frac{\varepsilon_0 \varepsilon_r w}{h} \, \mathrm{F\,m^{-1}}$$

$$= \frac{8.854 \times 10^{-12} \times 5 \times 3 \times 10^{-3}}{2 \times 10^{-3}} \, \mathrm{F\,m^{-1}}$$

$$= 66 \, \mathrm{pF\,m^{-1}}$$

$$= 0.66 \, \mathrm{pF\,m^{-1}}$$

2.9.6 ENERGY STORAGE

We can use capacitors as energy storage devices – on computer memory boards, charged capacitors can supply power to the memory chips if the main supply fails. So, we can use a capacitor to store energy, but how and where does a capacitor store the energy?

As we have already seen, if the capacitor is holding a charge an electric field exists in the dielectric. When the capacitor discharges through an external circuit, charges appear to move across the dielectric, against the E field. As they move against the field, work is done and energy is lost. When a discharged capacitor is connected to a voltage source, the reverse takes place.

To find the stored energy, let us take a capacitor connected to a source of V volts. If we increase the voltage by dV, the stored charge will increase by dQ. If these changes occur in a time dt, the instantaneous current will be

$$i = \frac{\mathrm{d}Q}{\mathrm{d}t}$$

$$= C\frac{\mathrm{d}V}{\mathrm{d}t} \tag{2.50}$$

The instantaneous power is given by

$$iV = CV\frac{dV}{dt}$$

and so the energy supplied in raising the voltage from V to $V + dV$ in time dt is

$$iV\ dt = CV\frac{dV}{dt} \times dt$$

or

$$energy = CVdV$$

Thus, the total energy supplied in raising the capacitor voltage from zero to V is

$$energy = \int_0^V CVdV$$

$$= C\left.\frac{V^2}{2}\right|_0^V$$

$$= \frac{1}{2}CV^2\ \text{J} \qquad (2.51)$$

Let us now find where the energy is stored. If we substitute for the capacitance in Equation (2.51), we get

$$energy = \frac{1}{2}\frac{\varepsilon_0\varepsilon_r \times area}{d}V^2$$

$$= \frac{1}{2}\varepsilon_0\varepsilon_r\frac{V^2}{d} \times area$$

$$= \frac{1}{2}\varepsilon_0\varepsilon_r E\ V \times area \qquad (2.52)$$

We can find the energy per unit volume by dividing this equation by the volume of the capacitor. So,

$$energy = \frac{1}{2}\varepsilon_0\varepsilon_r E\frac{V}{d} \times \frac{area}{area}$$

$$= \frac{1}{2}\ DE\ \text{Jm}^{-3} \qquad (2.53)$$

So, from Equation (2.53) it would appear that the electrostatic field stores the energy and not the capacitor plates. Such a point of view is quite useful if we consider fields in free-space in Chapter 9.

Example 2.13

A 10 μF capacitor has a potential difference between the plates of 50 V. Determine the energy stored in the electrostatic field.

Solution

As the capacitance is quoted, we can use Equation (2.51) to give

$$\text{energy} = \frac{1}{2}CV^2$$

$$= \frac{1}{2} \times 10 \times 10^{-6} \times 50^2$$

$$= 12.5\,\text{mJ}$$

2.9.7 FORCE BETWEEN CHARGED PLATES

We have already seen that charges exert a force on each other. We have also seen that charged circular plates radiate electric flux in a cylinder. We should therefore expect that two charged plates will exert a force on each other. To find this force, we could adapt our model of flux distribution in a parallel plate capacitor, and then calculate the force between the plates. However, there is a simpler method.

Let us consider the parallel plate capacitor shown in Figure 2.22. This capacitor stores a certain amount of energy given by

$$\text{energy} = \frac{1}{2}DE \times \text{area} \times l$$

Now, there will be a force of attraction between the two plates. If we move the top plate by a small amount dl, we do work against the attractive force. As we are moving the top plate a very small amount, the E field will hardly vary. As the work done must equal the change in stored energy, we can write

$$Fdl = \frac{1}{2}DE \times \text{area} \times (l + dl) - \frac{1}{2}DE \times \text{area} \times l$$

$$= \frac{1}{2}DE \times \text{area} \times dl \qquad (2.54)$$

As D is the flux density, Equation (2.54) becomes

$$F = \frac{1}{2}\psi E$$

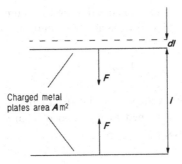

FIGURE 2.22 Force between two charged plates.

or

$$F = \frac{1}{2}QE \text{ N} \qquad (2.55)$$

where we have made use of Gauss' law. Depending on the particular application, the force between the two plates can be quite high.

Example 2.14

An electrostatic voltmeter consists of two square plates, of area 25 cm², separated by a distance of 2 cm in air. One of the plates is fixed, while the other is attached to a spring mechanism that deflects a needle in front of a calibrated scale. The constant of proportionality for the meter is 10° per 1 μN of force. Determine the angular displacement of the needle if a potential of 500 V is maintained across the plates.

Solution

The force between the two plates is given by Equation (2.55) as

$$F = \frac{1}{2}QE \text{ N}$$

Thus,

$$F = \frac{1}{2} \times \text{capacitance} \times 500 \times \frac{500}{2 \times 10^{-2}}$$

$$= 7 \times 10^{-6} \text{ N}$$

This corresponds to an angular displacement of 70°.

2.9.8 LOW-FREQUENCY EFFECTS AND DISPLACEMENT CURRENT

So far we have only considered the effects of direct current (d.c.) on capacitors. However, in electrical engineering we usually find capacitors in alternating current (a.c.) circuits. So, what effect does a capacitor have on a.c. signals?

Figure 2.23 shows a capacitor connected to an a.c. source. The voltage across the capacitor varies with time and so, if we assume the source to be sinusoidal, we can write

$$v_s(t) = V_{pk} \sin \omega t \qquad (2.56)$$

where V_{pk} is the peak source voltage, and ω is the angular frequency of the source.

Now, the capacitance is defined as the ratio of charge to potential difference between the capacitor plates, i.e.,

$$C = \frac{Q}{V}$$

or

$$Q = CV \qquad (2.57)$$

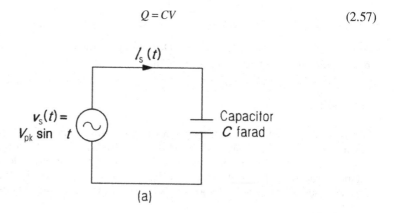

(a)

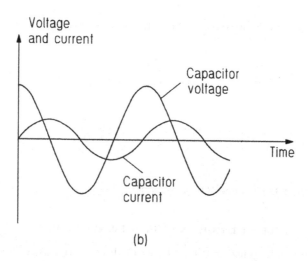

(b)

FIGURE 2.23 (a) Capacitor connected to an a.c. source and (b) relationship between capacitor voltage and current.

As the source is varying with time, we can write

$$q(t) = Cv_s(t)$$

$$= C\,V_{pk}\sin\omega t$$

If we differentiate this equation with respect to time, we get

$$\frac{d}{dt}q(t) = C\,V_{pk}\omega\cos\omega t$$

As current is the rate of change of charge with respect to time, we get

$$i_s(t) = \frac{d}{dt}q(t)$$

$$= C\,V_{pk}\omega\cos\omega t$$

$$= C\,V_{pk}\omega\sin(\omega t + 90°) \qquad (2.58)$$

So, when connected to an alternating source, the capacitor allows a current to flow, with the current leading the voltage by 90°. (Figure 2.23b shows the relationship between capacitor voltage and current.) We should note that the current flow is directly proportional to the angular frequency of the source, i.e., the higher the frequency, the larger the current flow. We can formalize this observation by defining the reactance of the capacitance, X_C, as

$$X_C = \frac{1}{\omega C}$$

$$= \frac{1}{2\pi f C} \qquad (2.59)$$

By combining this result with Equation (2.58) we get, after some rearranging,

$$i_s(t) = \frac{v_s(t)}{dt}\underline{/90°}$$

This is remarkably similar to Ohm's law, except that we are dealing with a.c. quantities, and there is a 90° phase shift involved.

So, when a capacitor is connected to an a.c. source, it provides a low resistance path for a.c. signals. Of course, if there is a d.c. voltage across the capacitor, no current will flow (if the capacitor is ideal). This makes capacitors very useful in smoothing power rails (where there might be some variation in supply voltage) and in connecting a.c. amplifier stages together – capacitors will let the a.c. signal through but block any d.c. levels.

Although this 'circuits' model indicates that a current will flow through a capacitor connected to an a.c. supply, it does not explain how the current gets from one plate to the other. Indeed, if the capacitor dielectric is ideal, there can be no flow of electrons from one plate to the other. So, how does the current magically cross the dielectric? To explain this we must use a field theory approach.

Figure 2.24a shows the state of the capacitor on positive half-cycles of the supply voltage. As can be seen, positive charge has built up on the lower plate. As we saw earlier in this section, these charges will radiate flux in upward and downward directions. The flux that radiates upwards will tend to attract negative charges to the top plate and repel any positive charges. Thus, in the positive half-cycle of the supply voltage, positive charges on the lower plate induce negative charges on the upper plate.

Let us now see what happens when the supply voltage has a negative half-cycle. Negative charges on the bottom plate will attract positive charges to the top plate and repel any negative charges. Thus, in the negative half-cycle of the supply voltage, negative charges on the lower plate induce positive charges on the upper plate.

This study of charge build-up shows why there is a phase difference between the supply voltage and the current – positive charges on one plate induce negative charges on the other plate, and vice versa. It also shows that, although charges appear to flow across the dielectric, they do not in reality – it is the electric flux that flows through the dielectric. If the supply voltage varies with time, the electric flux will also vary with time. Now, the units of electric flux are the same as charge, and so if we calculate the rate of change of flux with time, we will get units of coulomb per second, or amps, i.e.,

$$\text{rate of change of flux} = \frac{d\psi}{dt}$$

From Gauss' law we know that each unit charge radiates a unit of electric flux, i.e., $\psi = Q$. So, the rate of change of flux is the same as the rate of flow of charges in the wires connected to the capacitor, i.e.,

$$\text{rate of change of flux} = \frac{dQ}{dt}$$

$$= \text{capacitor current}$$

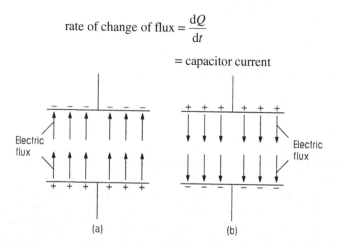

(a)   (b)

FIGURE 2.24   (a) Charge distribution on positive half-cycles and (b) charge distribution on negative half-cycles.

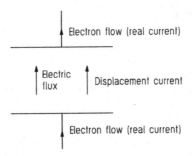

FIGURE 2.25   Conversion of electron flow into displacement current.

As the units of $d\psi/dt$ are the same as for current, we could regard $d\psi/dt$ as a current. In view of this, it is known as displacement current because it displaces charges from the opposite plate of the capacitor.

So, this 'fields' model of a capacitor has explained why there is a phase shift between the supply voltage and the current. It has also introduced us to the idea of displacement current. This is very important when considering electromagnetic radiation, or radio waves, and it is essential that readers are happy with the concept. Figure 2.25 summarizes the mechanism by which current apparently flows through a capacitor.

Example 2.15

A 22 µF capacitor is connected to a 4 V a.c. supply which has a frequency of 100 Hz. Determine the current taken from the supply. In addition, calculate the displacement current in the capacitor dielectric.

Solution

The frequency of the supply is 100 Hz, and so the angular frequency is

$$\omega = 2\pi f$$

$$= 2\pi \times 100$$

$$= 200\pi \text{ rad s}^{-1}$$

Now the reactance of the capacitor is given by

$$X_C = \frac{1}{\omega C}$$

and so,

$$X_C = \frac{1}{200\pi \times 22 \times 10^{-6}}$$

$$= 72.34 \ \Omega$$

Thus, the supply current is

$$i_s = \frac{V_s}{X_C}$$

$$= \frac{4}{72.34}$$

$$= 55.3 \text{ mA}$$

As the 'real' current changes to displacement current when it encounters the dielectric, we can write

$$\text{displacement current} = \frac{dQ}{dt}$$

$$= 55.3 \text{ mA}$$

Of course, we can never get an ideal dielectric. Thus, there will also be some flow of charge across the capacitor, i.e., there will be some 'real' current. We will meet this again in Section 4.4.

2.9.9 CAPACITANCE AS RESISTANCE TO FLUX

When considering the production of fields in a capacitor, it is sometimes useful to regard the capacitance as the resistance to the flow of electric flux. We have already seen, in Equation (2.35) that

$$Q = CV \tag{2.61}$$

and so, by applying Gauss' law,

$$\psi = CV \tag{2.62}$$

or

$$V = \psi \frac{1}{C} \tag{2.63}$$

Equation (2.63) relates the potential across the capacitors to the electric flux by way of the inverse of the capacitance. So, the higher the capacitance of a conductor system, the easier it is to produce electric flux. Thus, we can regard the capacitance as a measure of the resistance to electric flux. (Although some readers may be wondering why this point is being stressed, all will become clear when we compare electrostatics, electromagnetism and electroconduction in Chapter 5.)

Example 2.16

A 10 μF capacitor has a potential of 100 V d.c. across its terminals. Determine the flux through the capacitor. If the capacitance increases to 50 μF, determine the new flux.

Solution

We have a 10 μF capacitor with a 100 V across it. Thus, the flux through the capacitor is (Equation (2.62)),

$$\psi = CV$$

$$= 10 \times 10^{-6} \times 100$$

$$= 1 \, \text{mC}$$

The capacitance is now increased to 50 μF, and so the new flux is

$$\psi = CV$$

$$= 50 \times 10^{-6} \times 100$$

$$= 5 \, \text{mC}$$

So, by increasing the capacitance we have increased the flux through the capacitor. These fluxes are, of course, equal to the charge stored in the capacitor.

2.9.10 COMBINATIONS OF CAPACITORS

In engineering, we often have to make something new out of existing components. This is because manufacturers like to produce standard components to keep costs down. When designing a circuit we often need a particular value of capacitor that is not available from any source. We could pick a value close to the one we require, and then physically alter the area of the capacitor to get the capacitance required by filling the top down. This requires the use of a capacitance meter, a file, a steady hand and a great deal of patience! Although this can be done, it is not a very practical way to get a non-standard value of capacitance. Instead, we can produce non-standard capacitance values by combining standard values in parallel or series.

Figure 2.26a shows two capacitors, C_1 and C_2 in parallel. We require to find the equivalent capacitance of this arrangement. Let us connect a d.c. source, V_s, to the capacitors. Now both capacitors will have the same voltage across them, but store different charge. Thus, we can write

$$Q_1 = C_1 V_s \tag{2.64a}$$

and

$$Q_2 = C_2 V_s \tag{2.64b}$$

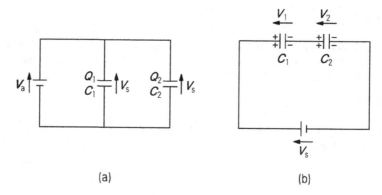

(a)   (b)

FIGURE 2.26   (a) Parallel connection of capacitors and (b) series connection of capacitors.

If we replace the two capacitors by a single equivalent one of value C_t, the charge on the new capacitor, Q_t, must be the same as on the parallel combination. Thus,

$$Q_t = C_t V_s \tag{2.65}$$

As Q_t must be the sum of the individual charges, in order to be equivalent, we can write

$$Q_t = Q_1 + Q_2$$

or

$$C_t V_s = C_1 V_s + C_2 V_s$$

Thus,

$$C_t = C_1 + C_2 \tag{2.66}$$

So, we can increase capacitance by adding another capacitor in parallel with the original.

Let us now consider a series combination of capacitors – Figure 2.26b. As before, we will connect this combination to a d.c. source and replace the capacitors with an equivalent one that will hold the same charge.

Now, let us assume that, when connected to the supply, a positive charge builds up on the left-hand plate of C_1. This charge induces an equal negative charge on the right-hand plate of C_1. The negative charge on this plate has to come from the left-hand plate of C_2, which leaves this plate positively charged. This positive charge on the left-hand plate induces a negative charge on the right-hand plate of C_2. As no charge leaves the circuit, the two capacitors store the same charge, but have different voltages across them. So, we can write

$$Q_1 = Q_2$$

or

$$C_1 V_1 = C_2 V_2 \qquad (2.67)$$

We are seeking to replace this series combination by a single capacitor of value C_t which must store a charge of Q_t when connected to a supply of V_s. Also, as the supply voltage must equal the individual voltage drops around the circuit, we can write

$$V_t = V_1 + V_2$$

and so,

$$\frac{Q_t}{C_t} = \frac{Q_1}{C_1} + \frac{Q_2}{C_2}$$

The charge in the circuit is a constant given by

$$Q_t = Q_1 = Q_2$$

and so,

$$\frac{Q_t}{C_t} = \frac{Q_1}{C_1} + \frac{Q_2}{C_2}$$

which gives

$$\frac{1}{C_t} = \frac{1}{C_1} + \frac{1}{C_2} \qquad (2.68)$$

So, we can decrease the capacitance by adding another capacitor in series with the original.

Example 2.17

A 10 µF is connected in a circuit. What is the effect of placing a 1 µF capacitor in parallel with it? What happens if the 1 µF is connected in series with the original?

Solution

We have a 10 µF capacitor, and a 1 µF connected in parallel. So, the total capacitance is

$$C_t = 10 + 1$$

$$= 11 \mu F$$

which indicates that the capacitance has barely altered.

If we now connect the 1 μF capacitor in series, we get a new capacitance of

$$\frac{1}{C_t} = \frac{1}{10} + \frac{1}{1}$$

$$= 0.1 + 1$$

$$= 1.1$$

i.e.,

$$C_t = 0.91 \mu F$$

So, the 1 μF capacitor dominates the 10 μF capacitor when it is in series with the larger capacitor.

2.10   SOME APPLICATIONS

We will look at two applications of electrostatics: the electron gun and the electrostatic precipitator. Electron guns are often found in high-school physics experiments to study diffraction of electrons amongst other experiments.

Figure 2.27a is a cathode ray tube, CRT, that uses an electron gun (Figure 2.27b) to produce an image on the screen. (Although such technology has now been largely superseded by LCD and plasma screens, electron guns are still used in physics experiments to study electrons.) The electron gun is capable of accelerating electrons to speeds approaching that speed of light. At the very rear of the device is a heater that causes the cathode to emit electrons. (These electrons are thermally excited because the cathode reaches temperatures in excess of 1800°C. This is thermionic emission.) Of course, if the gun contains air, the electrons will lose momentum and may fail to reach the target. Thus, it is important that the gun has a vacuum inside and this is usually done by burning magnesium in the tube after it has been sealed.

Now for some simple calculations, let us assume that the anode is grounded and that the cathode is at a negative voltage of V volt. Let us also assume that the distance between the anode and cathode is x metre. When the cathode produces electrons by thermionic emission, they are accelerated by the E field. So, the force acting on one electron is

$$\boldsymbol{F} = q\boldsymbol{E} \tag{2.69}$$

This must be equal to the mass of the electron, m, times the acceleration due to the field. Thus,

$$\boldsymbol{F} = q\boldsymbol{E} = m\boldsymbol{a}$$

giving

$$\boldsymbol{a} = \frac{q}{m}\boldsymbol{E} \tag{2.70}$$

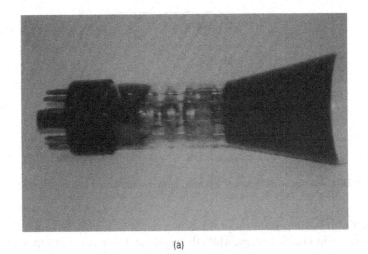

(a)

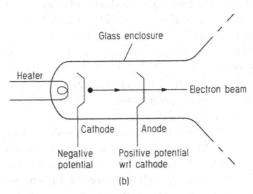

Glass enclosure

Heater

Electron beam

Cathode   Anode

Negative   Positive potential
potential   wrt cathode

(b)

FIGURE 2.27 (a) A typical cathode ray tube (CRT) and (b) basic structure of the electron gun.

If we assume that the electrons are initially at rest, we can find their final velocity using

$$v^2 = 2\frac{q}{m}E\,x$$

$$= 2\frac{q}{m}\frac{V}{x}\,x$$

$$= 2\frac{q}{m}V$$

We can rearrange this equation to give

$$\frac{1}{2}mv^2 = qV \qquad\qquad (2.71)$$

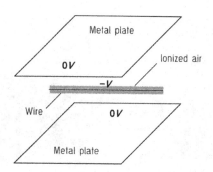

FIGURE 2.28   Schematic diagram of a typical electrostatic precipitator.

Equation (2.71) is simply a form of the conservation of energy – the left-hand side is the increase in kinetic energy, while the right-hand side is the electron energy in electron-volts. As an example, let us take a cathode voltage of −20 kV. As the mass of an electron is 9.1×10^{-31} kg, this gives a final velocity of 8.4×10^7 m s^{-1} – quite fast!

We can also find electrostatics at work in industry. The air in a factory often contains oil and dust particles. Although some form of vacuum cleaner can be used, the maintenance costs are quite high. An alternative is to ionize the particles and use electrostatics to attract them to charged metal plates where they can be removed. Such a device is the electrostatic precipitator, and Figure 2.28 shows the schematic diagram of a typical example.

As can be seen from the figure, the precipitator basically consists of a wire placed between two metal plates. In operation, the wire is maintained at a large negative potential, typically −50 kV, with respect to the plates. Under these conditions, the *E* field close to the wire is large enough to ionize the air. The negative field repels the freed electrons, while the wire attracts the positive charges. So, the electrons are accelerated towards the outer plates.

Now, if there are dust particles between the plates, the free electrons will attach themselves to the dust, so making them negatively charged. Thus, the positively charged plates attract the negatively charged particles, so removing them from the atmosphere.

The precise analysis of this situation is complicated because the plates are not coaxial to the wire. In this case, we must plot the electrostatic field strength between the plates to find out what happens to the dust particles. Under such circumstances, it is probably easier to build a prototype and experiment.

2.11   SUMMARY

We started this chapter by examining a fundamental law relating to the force between point charges – Coulomb's law. We then went on to develop the ideas of electric flux, electric field and potential. Again, we were only concerned with point charges. The relevant formulae are summarized here:

$$F = \frac{q_1 q_2}{4\pi\varepsilon r^2} r \qquad (2.72)$$

$$D = \frac{q_1}{4\pi r^2} r \qquad (2.73)$$

$$E = \frac{q_1}{4\pi \varepsilon r^2} r \qquad (2.74)$$

$$V = \frac{q_1}{4\pi \varepsilon r} \qquad (2.75)$$

We then considered certain charge distributions, line charges, surface charges and volume charges. In all of these cases, we were able to apply Gauss' law to simplify the analysis.

We next examined various types of capacitor: parallel plate, coaxial cable, twin feeder, wire over a ground and microstrip lines. In all cases, we chose to ignore the effects of the field at the edges of the conductors. The capacitances are reproduced here:

Parallel plate
$$C = \frac{\varepsilon_0 \varepsilon_r \times \text{area}}{d} \, \text{F} \qquad (2.76)$$

Coaxial cable
$$C' = \frac{2\pi \varepsilon_0 \varepsilon_r}{\ln(b/a)} \, \text{F m}^{-1} \qquad (2.77)$$

Twin feeder
$$C' = \frac{\pi \varepsilon}{\ln\big((d-a)/a\big)} \, \text{F m}^{-1} \qquad (2.78)$$

Wire over ground
$$C' = \frac{2\pi \varepsilon_0}{\ln\big((d/2-a)/a\big)} \, \text{F m}^{-1} \qquad (2.79)$$

Microstrip
$$C' = \frac{\varepsilon_0 \varepsilon_r w}{h} \, \text{F m}^{-1} \qquad (2.80)$$

or
$$C' = \frac{2\pi \varepsilon_0 \varepsilon_r \times \text{area}}{\ln(h/w)} \, \text{F m}^{-1} \qquad (2.81)$$

We also encountered the fundamental formula:

$$Q = CV \qquad (2.82)$$

We then went on to consider the storage of energy by a capacitor. We found that the energy can be regarded as either stored on the capacitor plates or stored in the electrostatic field between the plates. This is an important concept to grasp as it shows the equivalence between a field theory approach and the more familiar 'circuits' approach. The stored energy is given by

$$\text{energy} = \frac{1}{2}CV^2 \text{ J} \tag{2.83}$$

or

$$\text{energy} = \frac{1}{2}DE \text{ J m}^{-2} \tag{2.84}$$

The question of how the current 'flowed' through an ideal capacitor was then examined. This introduced us to the idea of displacement current and showed us that the 'real' current converted to 'displacement' current and back again as it crossed the dielectric. This is a very important point, which we cannot explain by the 'circuits' approach. Indeed, we found that we can regard the capacitance as the 'resistance' to the flow of flux. This acts as a link between the fields and circuits approaches.

We also saw that the reactance of a capacitor is given by

$$X_C = \frac{1}{2\pi f C} \tag{2.85}$$

with the capacitor current leading the supply voltage by 90°.

Next, we examined parallel and series combinations of capacitors. We saw that capacitance can be increased by adding another capacitor in parallel with the original and decreased by adding series capacitance.

We concluded this chapter with a brief examination of two applications of electrostatics: electron acceleration in an electron gun and electrostatic precipitation. In the first example, we saw that electrons can be accelerated to very high velocities by a potential difference of, typically, 20 kV. Electrostatic precipitators also use potentials of this order to attract dust and smoke particles to metal plates, so cleaning the air.

3 Electromagnetic Fields

When we considered electrostatics in the previous chapter, we started with the force between isolated point charges. This introduced us to the ideas of flux density and electric field strength. Now that we are considering magnetism, we can also start at the same point – isolated north or south monopoles – and Section 3.1 develops some basic ideas based around this concept. However, no one has yet found isolated magnetic poles and so we will quickly encounter the magnetic field generated by a current-carrying elemental wire in Section 3.2 – hence the term electromagnetism. Once we have grasped this idea, we will leave magnetic monopoles behind. (Of course, if someone does find magnetic monopoles, we will have to rewrite all the textbooks – this one included!)

3.1 SOME FUNDAMENTAL IDEAS

At about the same time that Coulomb was examining the force between isolated charges, he was also experimenting with magnetism (1785). In common with electrostatics, he found that the force between two magnetic poles decreases as the inverse of the square of the distance separating them, i.e.,

$$F = \frac{p_1 p_2}{kr^2} r \qquad (3.1)$$

where
  F is the vector force between the two poles (N)
  p_1 and p_2 are the strengths of the magnetic poles (Wb)
  k is a constant of proportionality
  r is the distance between two poles (m) and
  r is the unit vector acting in the direction of the line joining the two charges

This is the exact parallel of Coulomb's law as applied to isolated point charges. The force is repulsive if the poles are alike and attractive if the poles are dissimilar (Figure 3.1).
  We can extract a factor of 4π from the constant k to give

$$F = \frac{p_1 p_2}{4\pi \mu r^2} r \qquad (3.2)$$

where μ is the permeability – a material property. If we use the SI system of units, the force is in Newton if the pole strengths are in Weber (named after Wilhelm Eduard Weber, 1804–1891, the German physicist noted for his study of terrestrial magnetism), μ is in H m^{-1} and r is in metre. The reason for the choice of units for μ will become clear when we consider inductance in Section 3.11.

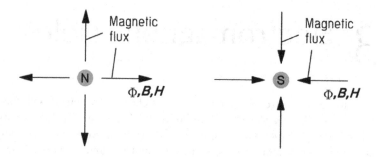

FIGURE 3.1   Two separate magnetic monopoles in free-space.

It is now a simple matter to introduce the idea of magnetic field strength in the same way that we introduced electric field strength. The force on the pole p_2 is

$$F = p_2 H \tag{3.3}$$

where H is the magnetic field strength due to pole p_1 given by

$$H = \frac{p_1}{4\pi\mu r^2} r \tag{3.4}$$

with units of N W^{-1}.

If we adapt Gauss' law to magnetostatics, we can say that the magnetic flux emitted by a pole is equal to the strength of the pole. Thus, we can define the magnetic flux density as

$$B = \frac{p_1}{4\pi r^2} r \tag{3.5}$$

with units of Wb m^{-2}.

We can combine Equations (3.4) and (3.5) to give

$$B = \mu H \tag{3.6}$$

In common with electrostatics, the value of the constant of proportionality (in this case the permeability) is dependent on the material. When working with magnetism, it is common practice to work with the permeability relative to free-space. Thus,

$$\mu_r = \frac{\mu}{\mu_0}$$

where μ_0 is the permeability of free-space with value $4\pi \times 10^{-7}$ H m^{-1}. Unfortunately, the relative permeability of a magnetic material varies according to the flux density so we shouldn't really refer to it as a material constant. (We will meet the relative permeability again when we consider magnetic materials in Chapter 7.)

So, we have developed a model in which magnetic flux emanates from an isolated pole (assumed to be a point source) in a radial direction. This model is identical to that adopted for isolated point charges in Chapter 2. However, we must use caution from this point onwards as no one has yet found isolated magnetic monopoles. Instead, we must adapt our model to current-carrying wires.

Example 3.1

A single 10 µWb magnetic monopole is situated in air. Calculate the magnetic field strength at a distance of 0.5 m from the monopole. In addition, find the flux density and the force on an identical monopole at the same distance.

Solution

We have a single monopole of strength 10 µWb situated in air. Now, the magnetic field strength is (Equation (3.4))

$$H = \frac{p_1}{4\pi\mu r^2}\, r$$

and so,

$$H = \frac{10 \times 10^{-6}}{4\pi \times 4\pi \times 10^{-7} \times 0.5^2}\, r$$

$$= 2.53r \text{ N Wb}^{-1} \text{ in a radial direction}$$

From Equation (3.6), we have
$B = \mu H$
And so,

$$B = 4\pi \times 10^{-7} \times 2.53r$$

$$= 3.2 \times 10^{-6}r$$

$$= 3.2r\,\mu\text{Wb m}^{-2} \text{ in a radial direction}$$

If we place an identical monopole at 0.5 m from the original, the force on this monopole is

$$F = p_2 \cdot H$$

$$= 10 \times 10^{-6} \times 2.53r$$

$$= 2.53r\,\mu\text{N (repulsive)}$$

So, even if we have low-strength monopoles, the field strength can be quite high (2.53 N Wb⁻¹). However, even with such a high value of H, the flux density is low (3.2 µWb m⁻²). These values are quite typical when considering magnetism – the magnetic field strength can be quite high, but the flux density will be low.

(Particle accelerators and fusion reactors use exceptionally high flux densities, of values greater than 1 Wb m⁻².)

3.2 SOME ELEMENTARY CONVENTIONS USED IN ELECTROMAGNETISM

Having established that the inverse square law applies to isolated magnetic monopoles, we will now examine the magnetic field produced by a current-carrying conductor. (Although readers may not be very familiar with this effect, everyone has come across it. Any piece of rotating electrical equipment – power tools, alternators, starter motors, etc., – relies on the magnetic fields produced by a current-carrying wire. We will consider a simple motor/generator in Chapter 7.)

In 1819, Hans Christian Oersted (a Danish physicist) demonstrated that a magnetic field surrounds a current-carrying wire. This was a very important discovery because it unified the separate sciences of electricity and magnetism into one science – electromagnetism. (Indeed, this was the first indication that two different forces of Nature could be unified. The search is now on for a Grand Unified Theory that would explain life, the Universe and everything!)

Oersted plotted the field surrounding a current-carrying wire using a compass. As Figure 3.2 shows, the magnetic field is coaxial to the wire. The direction of the field depends on whether the current flows up, as in Figure 3.2a, or down the wire, as in Figure 3.2b.

We now come across some conventions:

1. **A cross denotes current flowing into the page. A dot denotes current flowing out of the page.**

   Figure 3.2c shows these two conventions.

   Readers who play darts might like to imagine a dart thrown into the page. As the dart travels away from us, we see the cross of the feathers. Thus, a cross denotes current travelling away from us, down the wire. If the dart is travelling towards us, we see the point of the dart first – until it hits us! Hence, a dot corresponds to current travelling towards us, up the wire.

2. **The right-hand corkscrew rule gives the direction of the flux.**

   Figure 3.2c shows this rule.

   Most of us are familiar with the action of a corkscrew. We can use this action to determine the direction of the magnetic field. If we have current flowing away from us into the page, a corkscrew will have to turn in a clockwise direction to follow the current. Thus, the field acts in a clockwise direction. If the current is flowing towards us out of the page, the corkscrew acts in an anticlockwise direction. Thus, the field acts in an anticlockwise direction.

3. **Clockwise rotation of the field denotes a south pole. Anticlockwise rotation of the field gives a north pole.**

   We can easily apply this rule by drawing little arrows on the letter N (for north) and S (for south) as Figure 3.2d shows.

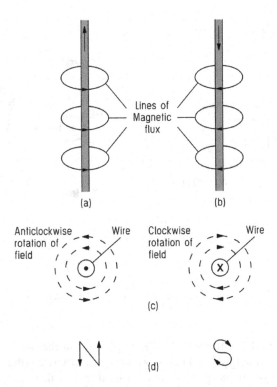

FIGURE 3.2 (a) Magnetic field produced by upward flowing current, (b) magnetic field produced by downward flowing current, (c) dot and cross conventions of current flow and (d) north and south pole conventions.

So, Oersted showed that a current-carrying wire generates a coaxial magnetic field. The field lines that we have drawn in Figure 3.2 are lines of magnetic flux. It is important to note that these flux lines act in a completely different direction to those we met in the previous section. This is a very important point to grasp when we consider the field surrounding a current-carrying wire, we cannot use the simple monopole model.

In the next section, we will examine the magnetic field produced by a simple current element – the Biot-Savart law.

3.3 THE BIOT-SAVART LAW

Let us consider the current-carrying conductor shown in Figure 3.3. Oersted showed that this conductor will generate a magnetic field that is coaxial to the wire. So, if we place an imaginary unit north pole at a point P, distance r from a small elemental section of the wire of length dl, the wire will experience a force that will tend to push it to the left. (The plan view in Figure 3.3b shows why this is so.) In addition, there will be an equal and opposite force on the north pole due to the field surrounding the wire.

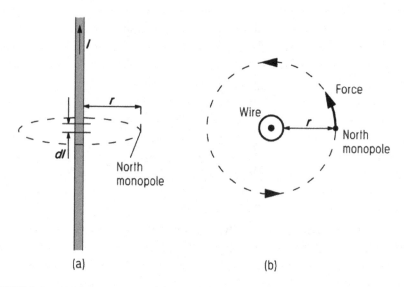

FIGURE 3.3 (a) Magnetic field produced by a current-carrying wire and (b) plan view of wire/magnetic field.

Let us try to find the magnetic field strength, δH_1, at the north pole, due to the current element formed by I and dl. As we have just discussed, the current element produces a force on the north pole, and the north pole will produce an equal and opposite force on the current element. If we can find these two forces, and then equate them, we should get an expression for the magnetic field strength generated by the wire.

Let us initially consider the field at dl due to the imaginary north pole of strength p_N. As this north pole is a point source, it emits magnetic flux in a radial direction. Thus, we can write the flux density as

$$B_N = \frac{p_N}{4\pi r^2}\, r \tag{3.8}$$

Direct experimental measurement shows that the force on a current-carrying conductor placed in a magnetic field is given by

$$F = BIl \tag{3.9}$$

where B is the flux density of the magnetic field in which the wire is placed, I is the current flowing through the wire and l is the length of the wire. (We can intuitively reason that this equation is correct by noting that powerful electric motors require a large electric current and contain a large amount of wire – they are very heavy!)

By combining Equations (3.8) and (3.9), we find that the force on the element dl due to the field emitted by the north pole is

$$dF = \frac{p_N}{4\pi r^2} I\, dl \tag{3.10}$$

Let us now turn our attention to the force on the north pole produced by the current element. The current element formed by the current I and the length dl produces a magnetic field strength of dH_1, at the north pole. By applying Equations (3.3) and (3.4), the force on the north pole due to the current element is

$$dF = p_N dH_1 \qquad (3.11)$$

By equating Equations (3.10) and (3.11), we get

$$\frac{p_N}{4\pi r^2} I \, dl = p_N dH_1$$

and so,

$$\mathbf{dH}_1 = \frac{I \, dl}{4\pi r^2} \boldsymbol{\phi} \qquad (3.12)$$

with direction into the page (ϕ is the unit vector acting into the page), this is the magnetic field strength due to the small current element formed by I and dl. To find the total field produced by the wire, we should integrate this equation with respect to the length. However, Equation (3.12) only gives the field at a point directly opposite the current element we are considering. To find the field we require, we must do some resolving of components.

Figure 3.4 shows the situation we now seek to analyze. If we draw a line from the point P to the current element, we find that it makes an angle to the current element

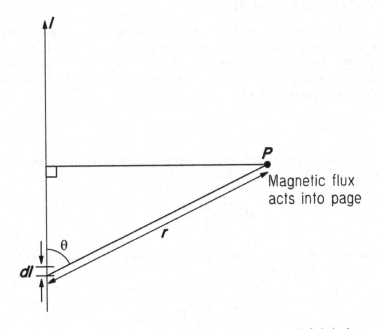

FIGURE 3.4 Construction for finding the magnetic field of an infinitely long current-carrying wire.

of θ. Now, if the wire is infinitely long, θ will vary from 0 to π and we can intuitively reason that at the two extremes of the wire, $+\infty$ and $-\infty$, the force on a north pole placed at the point P will be zero. So, if we modify Equation (3.12) by a factor $\sin \theta$, we will get

$$\mathbf{d}H = \frac{I\,\delta l}{4\pi r^2}\sin\theta \qquad\qquad (3.13)$$

with direction into the page. (We can check this equation by noting that $\sin 0$ and $\sin \pi$ are both zero. Thus, the force on the north pole generated by current elements at the extremes of the wires, at $+\infty$ and $-\infty$, is zero. The use of $\sin \theta$ in Equation (3.13) is deliberate: if we use vector algebra, the $\mathbf{H}$ field is proportional to the cross product of l and the vector δl drawn normal to the wire. The cross product introduces the factor $\sin \theta$.)

Although Equation (3.13) is known as the Biot-Savart law, we should really credit it to Ampère (1820) who originally did experiments on current-carrying conductors. (Biot and Savart were colleagues of Ampère.) In order for this equation to be of any practical use, we must integrate the current element over the length of our conducting wire. This we will do in Section 3.6.

We will now re-examine the idea of magnetic flux, magnetic flux density and magnetic field strength as produced by a current element.

3.4 ELECTROMAGNETIC FLUX, FLUX DENSITY AND FIELD STRENGTH

When we considered electrostatics in the previous chapter, we made use of Gauss' law to state that the electric flux emitted by a positive charge was equal to the enclosed charge. Now that we are considering electromagnetic fields, we can adapt Gauss' law to state that the magnetic flux generated by a current element is equal to the product of the current and element length. Thus,

$$p = I\,dl\,\text{Wb} \qquad\qquad (3.14)$$

As we have just seen, if we consider the isolated current element, $I\,dl$, of Figure 3.5, we have a fractional magnetic field strength, dH, of

$$\mathbf{d}H = \frac{I\,dl}{4\pi r^2}\sin\theta \quad \text{A}\,\text{m}^{-1} \qquad\qquad (3.15)$$

acting into the page. If we introduce a north pole at the point P, Equation (3.11) shows that it will experience a force of

$$\mathbf{d}F = \frac{I\,dl}{4\pi r^2}\sin\theta\ p_N \qquad\qquad (3.16)$$

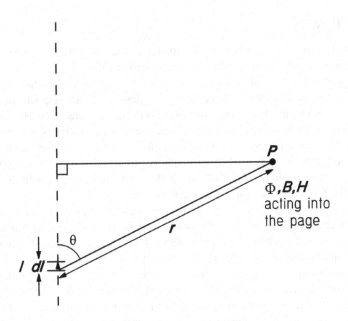

FIGURE 3.5 Magnetic field due to an isolated current element.

So, the force on a magnetic pole is directly dependent on the magnetic field strength produced by the wire. This agrees with our model of electrostatics. As $B = \mu H$, the fractional magnetic flux density is given by

$$dB = \mu \frac{I\ dl}{4\pi r^2} \sin\theta \qquad (3.17)$$

We should note that if we introduce a current element of strength p, at the point P, the force on this element will be given by

$$dF = \mu \frac{I\ dl}{4\pi r^2} \sin\theta\ p_i$$

or

$$dF = dBp_i \qquad (3.18)$$

Thus, the force on a current element is directly dependent on the magnetic flux density produced by the wire, and not the magnetic field strength. We must exercise great care here as this is often a source of confusion.

We have now completed our initial study of magnetostatics and electromagnetism. In the next section, we will compare the fundamental equations of electrostatics, magnetostatics and electromagnetism.

3.5   COMMENT

We can now compare our models of electrostatics, magnetostatics and electromagnetism. Table 3.1 lists the fundamental formulae we have met.

A glance at Table 3.1 shows a lot of similarity between electrostatics and magnetostatics. We should expect this because we developed our magnetostatic model along the lines of our electrostatic model. However, there are some obvious differences between electrostatics and electromagnetism: in electrostatics, the flux density is independent of any change in the surrounding material, whereas in electromagnetism, the flux density is directly proportional to a material property (the permeability). As we will see in Chapter 7, this is due to the ability of certain materials to increase the flux density produced by a coil of wire.

We can also make one further observation: in electrostatics, the force on an isolated point charge is dependent on the field strength. Similarly, in electromagnetism, the force on an isolated magnetic pole is dependent on the field strength, but the force on a current element depends on the flux density. This can be a source of confusion and we must exercise great care here. (This is because the science of magnetostatics developed in parallel with electrostatics, with isolated poles and charges being assumed in both cases. As we have seen, this means that the fundamental formulae for electrostatics and magnetostatics are very similar. With Oersted's discovery of electromagnetism, it was soon realized that it was the magnetic flux density and not the field strength produced by a current element that determines the force on a current-carrying wire. So, we now have electromagnetic terms that do not match the electrostatic counterparts. The situation might have been different if Oersted had discovered electromagnetism before magnetostatics had been formalized!)

In the following sections, we will apply the Biot-Savart law to a variety of current-carrying conductors. We begin by studying the field produced by a long current-carrying conductor. This will introduce us to Ampere's circuital law. (Although Ampere's circuital law came before the Biot-Savart law, the way our electromagnetic model has developed dictates this approach.)

TABLE 3.1
Comparison of Field Equations

	Force	Field Strength	Flux Density
Electrostatics	$F = \dfrac{q_1 q_2}{4\pi\varepsilon_0 r^2} = q_1 E$	$E = \dfrac{q_1}{4\pi\varepsilon_0 r^2}$	$D = \dfrac{q_1}{4\pi r^2}$
Magnetostatics	$F = \dfrac{p_1 p_2}{4\pi\mu_0 r^2} = p_2 H$	$H = \dfrac{p_1}{4\pi\mu_0 r^2}$	$B = \dfrac{p_1}{4\pi r^2}$
Electromagnetism	$dF = \dfrac{I\,dl}{4\pi r^2}\, p_2 \sin\theta = p_2\,dH$ (force on a magnetic pole) $dF = \mu_0 \dfrac{I\,dl}{4\pi r^2}\, p_1 \sin\theta = p_1\,dB$ (force on a current element)	$dH = \dfrac{I\,dl}{4\pi r^2}\sin\theta$	$dB = \mu_0 \dfrac{I\,dl}{4\pi r^2}\sin\theta$

3.6 MAGNETIC FIELD STRENGTH AND AMPERE'S CIRCUITAL LAW

Let us consider the current-carrying wire shown in Figure 3.6. We want to find the magnetic field strength at a point P, distance R from the wire. To find H at this point, we will determine the field strength due to a small current element, and then integrate the result over the length of the wire.

By applying the Biot-Savart law, we get

$$\mathbf{d}H = \frac{I\ dz}{4\pi r^2}\ \sin\theta \ \text{A m}^{-1} \tag{3.19}$$

acting into the page. Now, to find the total field strength, we need to integrate Equation (3.19) with respect to length. Unfortunately, as we move along the wire, the distance r and the angle θ will vary. So, we need to do some substitution and manipulation before we can do any integration.

Instead of working with the angle θ, we can simplify the integration if we use the angle α instead. So, with reference to Figure 3.6, we can see that $z = R \tan \alpha$ and so $dz = R\ d\alpha/\cos^2 \alpha$. As $\sin \theta = R/r = \cos \alpha$, we get $r = R/\cos \alpha$. Thus, Equation (3.19) becomes

$$\mathbf{d}H = \frac{1}{4\pi} \frac{R\ d\alpha}{\cos^2\alpha} \frac{\cos^2\alpha}{R^2} \cos\alpha$$

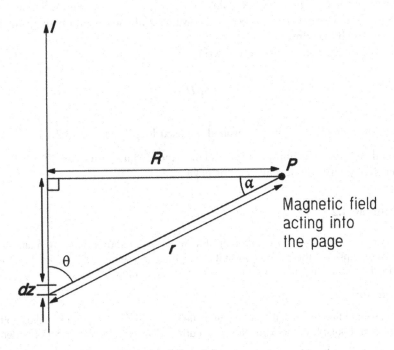

FIGURE 3.6   The magnetic field of an infinitely long current-carrying wire.

Now, as we move from $-\infty$ to $+\infty$ the angle α varies from $-\pi/2$ to $+\pi/2$. So,

$$H = \frac{I}{4\pi R} \int_{-\pi/2}^{+\pi/2} \cos^2\alpha \; d\alpha$$

$$= \frac{I}{4\pi R} \left| \sin\alpha \right|_{-\pi/2}^{+\pi/2}$$

$$= \frac{I}{4\pi R}(1+1)$$

and so,

$$H = \frac{I}{2\pi R} \text{ into the page.} \tag{3.20}$$

Let us take a moment to examine this equation in some detail. It shows that the magnetic field strength is proportional to the current in the wire. We should expect this because common sense tells us that a large magnetic field implies a large current. However, H is also inversely dependent on the term $2\pi R$. This is simply the circumference of a circle of radius R with the wire centre. We should also note that the field is independent of where exactly we are along the wire – provided we maintain a constant distance from the wire, the field is a constant. We can intuitively reason that these observations are correct because, as we have already seen, the magnetic field is coaxial with the wire.

In order to formalize this, we can write

$$I = \oint H \; dl \tag{3.21}$$

where $\oint$ denotes line integral around a closed loop. Equation (3.21) is Ampere's circuital law or, more simply, Ampere's law. (Andre Marie Ampere (1175–1836) was a French physicist who formulated this law.)

Example 3.2

A current of I A flows through a long straight wire, of radius a, situated in air. The current density in the wire is constant across the cross section of the wire. Plot the variation of H with radius both inside and outside of the wire.

Solution

We want to find the variation in magnetic field strength as a function of radius both inside and outside of the wire. Now, the current density in the wire is independent of radius, i.e, $I/\pi r^2$ is constant for $r < a$. so, the current enclosed by a circular path of radius r is

$$I' = \frac{I}{\pi a^2}\, \pi r^2 \quad r < a$$

Ampere's law, Equation (3.21), gives

$$I = \oint H \, dl$$

and so $I' = H_r 2\pi r$. Hence,

$$H_r = \frac{I'}{2\pi r}$$

$$= \frac{I}{\pi a^2}\, \pi r^2 \, \frac{1}{2\pi r}$$

$$= \frac{Ir}{2\pi a^2} \quad for\ r < a$$

i.e., the magnetic field is directly proportional to radius inside the wire.

Let us now consider a circular path outside the wire. This path encloses all the current, and so

$$I = H_r 2\pi r$$

Hence,

$$H_r = \frac{Ia}{2\pi r} \quad for\ r > a$$

i.e., the magnetic field is inversely proportional to radius outside the wire.

At the surface of the wire, these two values should be the same. So, as a check we can put $r = a$ to give

$$H_r = \frac{Ia}{2\pi a^2} = \frac{1}{2\pi a}$$

which is identical to the field outside the wire.

Figure 3.7 shows the variation of H with radius. If the wire is made of non-ferrous material, such as copper, the flux density will also follow the same variation.

3.7   THE FORCE BETWEEN CURRENT-CARRYING WIRES – THE DEFINITION OF THE AMPERE

We can now define the ampere. Readers may think that this is not worth considering since the ampere is simply a measure of current set down by international treaty. After all, we do not often have to concern ourselves with the definition of a metre.

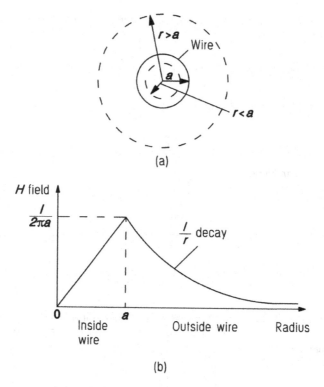

FIGURE 3.7   (a) Circular paths inside ($r<a$) and outside ($r>a$) a wire and (b) variation of H with radius.

However, as we will soon see, the definition of the ampere introduces us to the force between two current-carrying wires, and that is of some practical benefit.

Figure 3.8 shows the situation we are to analyze. Two current-carrying wires run parallel to each other, separated by a distance r. These wires each carry a current of I amperes. As we have already seen, current-carrying wires produce magnetic fields. As each wire carries the same current, the magnetic field produced by the left-hand conductor will exactly balance the field produced by the right-hand conductor at the point mid-way between the two conductors. Thus, the field at this point will be zero, resulting in the field distribution of Figure 3.8c. The weakening of the field between two wires shows that they attract each other.

Now, in Section 3.4 we met the force on an isolated north pole due to a current-carrying element. In this example, we do not have an isolated pole; instead we must consider a current element in the right-hand wire.

To find the force on the right-hand conductor, we need to find the magnetic flux density produced by the left-hand conductor. By applying Ampere's law, Equation (3.21), we can write

$$I = \oint H \, \mathrm{d}l$$

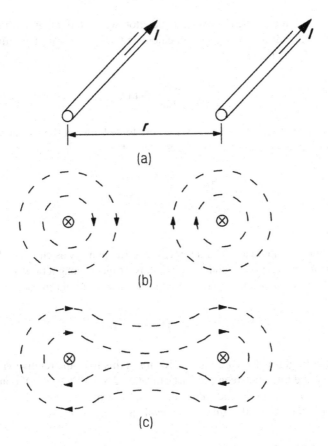

(a)

(b)

(c)

FIGURE 3.8 (a) Two parallel current-carrying wires, (b) magnetic field produced by each wire and (c) resultant field distribution.

and so $I = H_I 2\pi r$ or

$$H_I = \frac{I}{2\pi r} \qquad (3.22)$$

at the surface of the right-hand conductor. As $B = \mu_0 H$, the flux density at the right-hand conductor is

$$B_I = \frac{\mu_0 I}{2\pi r} \qquad (3.23)$$

Now, the force on a current element in the right-hand wire will be (Equation (3.9))

$$dF = B_1 I dl$$

$$= \frac{\mu_0 I}{2\pi r} I \, dl \qquad (3.24)$$

To find the total force on the right-hand conductor, we should integrate this equation with respect to length. However, we can calculate the force per unit length by dividing both sides by dl. Thus,

$$\frac{dF}{dl} = \frac{\mu_0 I^2}{2\pi r} \ \text{N}\,\text{m}^{-1} \tag{3.25}$$

If the separation between two wires is 1 m, the force between the two conductors is $2 \times 10^{-7}\,\text{N m}^{-1}$, and we use $4\pi \times 10^{-7}$ for μ_0, we get

$$\frac{dF}{dl} = 2 \times 10^{-7} = \frac{4\pi \times 10^{-7} \times I^2}{2\pi r \times 1}$$

and so,

$I = 1$ A

At this point we should ask why a force of $2 \times 10^{-7}\,\text{N m}^{-1}$ was chosen. The answer lies with the magnitude of μ_0 and ϵ_0. The study of electromagnetic waves in space (Chapter 9) shows that μ_0 and ε_0 are related to the speed of light in vacuum by

$$c = \frac{1}{\sqrt{\mu_0 \varepsilon_0}} \tag{3.26}$$

As the speed of light in a vacuum is $3 \times 10^8\,\text{m s}^{-1}$, this fixes the relative values of μ_0 and ϵ_0, and so the force between the wires must be $2 \times 10^{-7}\,\text{N m}^{-1}$. (Of course, if we changed our system of units, the values of μ_0 and ϵ_0 would change. However, the SI units are in use today, and this gives us our values of μ_0 and ε_0.)

3.8   THE MAGNETIC FIELD OF A CIRCULAR CURRENT ELEMENT

In electrical engineering, we often come across wound components – transformers and coils. Section 7.4 deals with transformers in detail. Here, we concern ourselves with the field produced by a circular piece of wire carrying a current. This will help us when we come to consider coils and solenoids in the next section.

Figure 3.9 shows a simple single-turn coil. We require to study the distribution of the magnetic field along (for simplicity) the axis of the coil. To analyze this situation, we will use the Biot-Savart law to find the field produced by a small section of the loop and then integrate around the loop to find the total field.

Let us consider a simple current element of length dl. Now, from the Biot-Savart law, the magnitude of the magnetic field strength at point P is given by

$$dH = \frac{I\ dl}{4\pi x^2}\sin\theta$$

As the angle θ is $\pi/2$ in this instance, we can write

$$dH = \frac{I\ dl}{4\pi x^2} \tag{3.27}$$

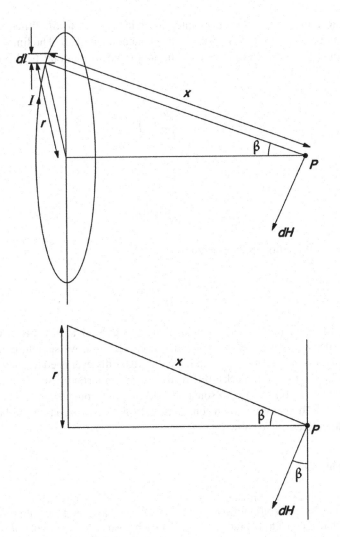

FIGURE 3.9 Construction to find the magnetic field of a single-turn coil.

Now, dH makes an angle to the horizontal of $\pi/2 - \beta$, and so we can resolve dH into vertical and horizontal components. When we integrate around the current loop, we find that the vertical component is zero due to the symmetry of the situation. (Interested readers can try this for themselves.) Thus, we need only consider the horizontal component of dH, i.e., dH_p. So,

$$\mathrm{d}H_p = \mathrm{d}H\,\cos\left(\pi/2 - \beta\right)$$

$$= \mathrm{d}H\sin\beta$$

$$= \frac{I\,\mathrm{d}l}{4\pi x^2}\sin\beta$$

To find the total field, we need to integrate with respect to dl. When we do the integration, the only thing that varies is the elemental length dl. The limits of dl will be zero and $2\pi r$ (the circumference of the loop) and so the total field produced at point P is

$$H_p = \frac{I \sin B}{4\pi x^2} \int_0^{2\pi r} dl$$

$$= \frac{I \sin B}{4\pi x^2} 2\pi r$$

$$= \frac{Ir}{2x^2} \sin \beta \qquad\qquad (3.28)$$

As $\sin \beta = r/x$, Equation (3.28) becomes

$$H_p = \frac{I \sin^3 B}{2r} \, \text{Am}^{-1} \qquad\qquad (3.29)$$

Although this equation gives the field along the axis of a single-turn coil, it is somewhat debatable whether it is of much use. After all, we seldom come across single-turn coils in real life, and even if we do, we hardly ever need to know what the field is. In spite of these remarks, this result is very important if we want to find the field due to a long coil of wire, or solenoid, and we often come across solenoids. This result is also used in physics when considering the field produced by Helmholtz coils, as the following example shows.

Example 3.3

A pair of coils, each consisting of N turns of wire and having a radius r metre, is situated in air. The coils face each other on the same axis, and they are separated by a distance equal to the radius of coils. If each coil carries a current of I A in the same direction, plot the variation in the magnetic field strength along the axis of the coils.

Solution

Figure 3.10a shows the situation we are to analyze. Now, to simplify the analysis, we will assume that the coils are very thin when compared with the distance between them. This means that we can assume the coils to be single turns carrying an effective current of NI amperes.

As we have just seen, the field along the axis of a single-turn coil is given by (Equation (3.29))

$$H = \frac{I \sin^3 \beta}{2r}$$

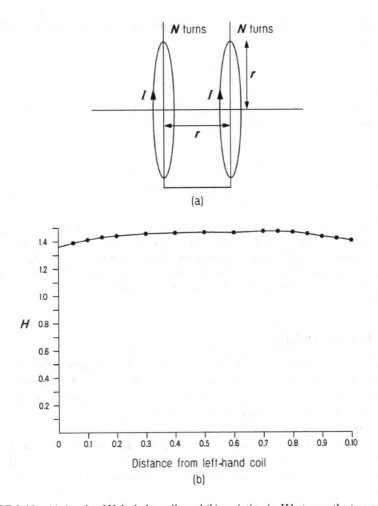

FIGURE 3.10 (a) A pair of Helmholtz coils and (b) variation in H between the two coils.

Thus, the field due to the left-hand coil is

$$H_1 = \frac{NI \sin^3 \beta}{2r}$$

Now, $\sin \beta = \dfrac{r}{\sqrt{(r^2 + x^2)}}$ and so,

$$H_1 = \frac{NI}{2r} \frac{r^3}{(r^2 + x^2)^{3/2}}$$

$$= \frac{NI}{2} \frac{r^2}{(r^2 + x^2)^{3/2}} \text{ from the left-hand coil.}$$

As the right-hand coil is identical to the left-hand one, it will produce a field strength of

$$H_1 = \frac{NI}{2} \frac{r^2}{\left(r^2 + (r-x)^2\right)^{3/2}}$$

Giving a total field strength of

$$H = H_l + H_r$$

$$= \frac{NI}{2} \frac{r^2}{\left(r^2 + x^2\right)^{3/2}} + \frac{NI}{2} \frac{r^2}{\left(r^2 + (r-x)^2\right)^{3/2}}$$

At this point, we could apply the binomial theorem to simplify this equation. However, the result is rather complicated. Instead, it is easier to plot the distribution of H directly, and Figure 3.10b shows the result. As this plot shows, the variation of H_l and hence B, is so small as to be neglected. Such coils are known as a Helmholtz pair (after Hermann von Helmholtz, 1821–1894, a German physicist).

3.9   THE SOLENOID

Figure 3.11 shows a long coil of wire, or solenoid. Such devices are often used as actuators with a bar magnet placed along the axis of the coil. Any current passing through the coil generates a magnetic field which forces the magnet in a particular direction. The magnet can then force a pair of contacts to close or push a lever to move something and so it is an actuator.

To determine the field at any point along the axis of the solenoid, we will consider an elemental section of the coil, of thickness dx, and calculate the field produced. We will then integrate along the length of the coil to find the total field produced.

Let us assume that the solenoid has N turns and a length of l metre. With these figures, the number of turns per unit length will be N/l. Now, if we take a small section of the coil, of length dl, the number of turns in this section will be $dl \times N/l$. By using the result from the last section, the magnetic field strength generated by this section of the solenoid is

$$dH_p = dl \frac{N}{l} \frac{I \sin^3 \beta}{2r} \tag{3.30}$$

acting along the axis of the coil.

We now need to integrate along the length of the solenoid. However, as we move along the axis, the angle β changes between the limits β_{max} and β_{min}. So, we have to substitute for dl in terms of β. As Figure 311b shows,

$$dl \sin \beta = d\beta \sqrt{r^2 + R^2}$$

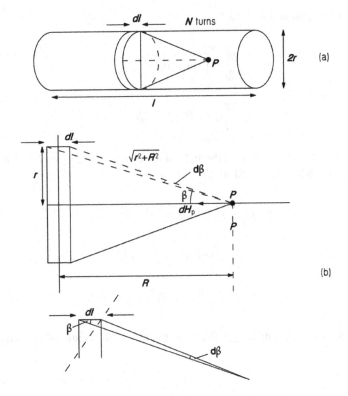

FIGURE 3.11 (a) A simple solenoid and (b) construction to determine the field produced by a solenoid.

and so,

$$dl = \frac{d\beta}{\sin \beta} \sqrt{r^2 + R^2}$$

Thus, Equation (3.30) becomes

$$dH_p = \frac{d\beta}{\sin \beta} \sqrt{r^2 + R^2} \frac{N}{l} \frac{I \sin^3 \beta}{2r}$$

$$= \frac{\sqrt{r^2 + R^2}}{2r} \frac{N}{l} I \sin^2 \beta \; d\beta \qquad (3.31)$$

Now, $\sin \beta = \dfrac{r}{\sqrt{r^2 + R^2}}$ and so we can write

$$dH_p = \frac{NI}{2l} \sin \beta \; d\beta$$

Thus, the total field at point P is

$$H_p = \frac{NI}{2l} \int_{\beta_{max}}^{\beta_{min}} \sin\beta \; d\beta$$

$$= \frac{NI}{2l}(\cos\beta_{min} - \cos\beta_{max}) \; A \; m^{-1} \tag{3.32}$$

We can use this equation to find the field at the centre of the solenoid. If P is at the centre, $\beta_{max} = 180° - \beta_{min}$, and so

$$H_p = \frac{NI}{2l}\left(\cos\beta_{min} - \cos(180° - \beta_{max})\right)$$

$$= \frac{NI}{2l}\left(\cos\beta_{min} + \cos\beta_{min}\right)$$

$$= \frac{NI}{l}\cos\beta_{min} \; A \; m^{-1} \tag{3.33}$$

If the solenoid is very long, $\beta_{min} \approx 0$, and so the field at the centre of the coil approximates to

$$H_p = \frac{NI}{l} A \; m^{-1} \tag{3.34}$$

Example 3.4

A 2-cm diameter solenoid is 10 cm long and has 30 turns per cm. When energized, the solenoid takes 2 A of current. Plot the variation in H and B along the axis of the coil. (Assume that the solenoid is air-cored.)

If a bar magnet of strength 10 mWb, and length 5 cm, is placed in the solenoid, plot the variation in force along the axis when the coil is energized.

Solution

We require to find the variation in H and B along the axis of the solenoid. Now Equation (3.32) gives the variation in H as

$$H_p = \frac{NI}{2l}(\cos\beta_{min} - \cos\beta_{max}) A \; m^{-1}$$

and so,

$$H_p = \frac{30 \times 10 \times 2}{2 \times 0.1}(\cos\beta_{min} - \cos\beta_{max})$$

$$= 3 \times 10^3(\cos\beta_{min} - \cos\beta_{max}) A \; m^{-1}$$

As the solenoid is air-cored, $\mu = 4\pi \times 10^{-7}$ and so

$$B_p = 3.8 \times 10^{-3} \left(\cos\beta_{min} - \cos\beta_{max} \right) \text{Wb m}^{-2}$$

Figure 3.12a shows the variation in H and B along the axis of the coil. As can be seen, the maximum field occurs at the centre of the solenoid.

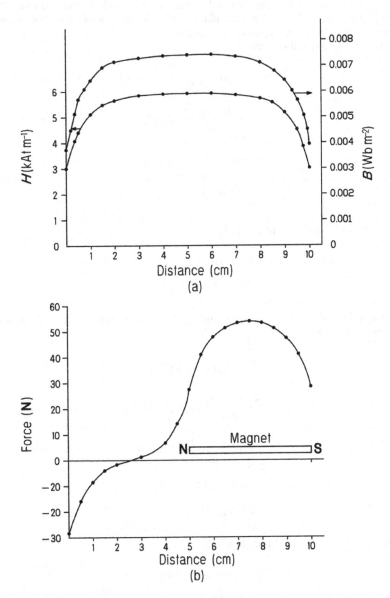

FIGURE 3.12 (a) Variation of B and H along the axis of a solenoid and (b) variation of force along the axis of a solenoid.

Now, if we place a 10 mWb bar magnet in the solenoid, the magnet will experience a force of

$$F = 10 \times 10^{-3} H_p$$

at each pole. As the magnet has a length of 5 cm, the two poles will experience different forces due to the variation in H as we move along the solenoid axis. Thus, it is a question of calculating the force on each pole and adding the two results to get the total force. As a sample calculation, we will take the left-hand pole of the bar magnet to be at the centre of the solenoid.

So, the H field at the centre is

$$H = \frac{NI}{l} \cos \beta_{min}$$

$$= \frac{30 \times 10 \times 2}{0.1} \cos \beta_{min}$$

$$= 6 \times 10^3 \times \frac{5}{\sqrt{5^2 + 1^2}}$$

$$= 5.9 \times 10^3 \text{ A m}^{-1}$$

Thus, the force on the left-hand pole is

$$F = 5.9 \times 10^3 \times 10 \times 10^{-3}$$

$$= 59 \text{ N}$$

The length of the magnet is 5 cm, and so the H field 5 cm to the right of the solenoid centre is

$$H = \frac{NI}{2l} (\cos \beta_{min} - \cos \beta_{max})$$

$$= \frac{30 \times 10 \times 2}{2 \times 0.1} (\cos \beta_{min} - \cos \beta_{max})$$

$$= 3 \times 10^3 \left\{ \frac{10}{\sqrt{10^2 + 1^2}} - \cos 90° \right\}$$

$$= 3 \times 10^3 \times 0.995$$

$$= 2.985 \times 10^3 \text{ A m}^{-1}$$

Thus, the force on the right-hand pole is

$$F = 2.985 \times 10^3 \times 10 \times 10^{-3}$$

$$= 29.85 \text{ N}$$

As the pole at this location is the opposite of the pole at the centre of the solenoid, the total force on the bar magnet is

$$F_t = 59 - 29.85$$

$$= 29.15\,\text{N}$$

Figure 3.12b shows the variation in this force as the magnet moves along the axis of the solenoid

3.10 THE TOROIDAL COIL, RELUCTANCE AND MAGNETIC POTENTIAL

Figure 3.13 shows the general form of a toroidal former which has a coil wound on it. This is basically a long solenoid which is bent so that the coil has no beginning or end. In practice, the formers used in toroidal coils are made of powdered ferrite which acts to concentrate the magnetic flux. Thus, leakage effects are minimal, so making the coil very efficient. This is put to good use in transformers, which we will encounter in Chapter 7. Here, we want to develop our model of electromagnetism further.

As we saw in the last section, the field at the centre of a long solenoid is given by (Equation (3.34))

$$H = \frac{NI}{l}\,\text{A m}^{-1} \tag{3.35}$$

where l is the length of the solenoid. As the coil is wound on a toroid, this field will be constant along the length of the coil. As the coil is circular, the length of the solenoid

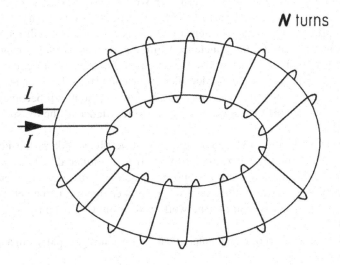

FIGURE 3.13 A typical toroidal coil.

will be the average circumference of the former. Now, as $B = \mu H$, the flux density in the former will be

$$B = \mu \frac{NI}{l} \text{ Wb m}^{-1} \tag{3.36}$$

with μ being the permeability of the former. As B is the flux density, Equation (3.36) becomes

$$\frac{\phi}{\text{area}} = \mu \frac{NI}{l}$$

and so,

$$NI = \phi \times \frac{1}{\mu \times \text{area}} \tag{3.37}$$

Let us examine this equation closely. The first term on the right-hand side of this equation is the magnetic flux that flows around the toroid. The second term is similar to our formula for the capacitance of a parallel plate capacitor, Equation (2.35). As we saw in Section 2.9, we can regard capacitance as a measure of the resistance to the flow of the flux. So, could we take this second term as a measure of resistance to magnetic flux?

Let us define the reluctance as

$$S = \frac{l}{\mu \times \text{area}} \tag{3.38}$$

with units of At Wb^{-1}, i.e., ampere-turns per Weber. (Most texts choose not to use units of turns as they are not part of the SI system. However, the use of ampere-turns is helpful when dealing with electromagnetism, and so we will retain them.) As Equation (3.38) shows, the reluctance is directly dependent on the length of the coil. Thus, if the toroidal former has a large circumference, the reluctance is high, and it becomes more difficult to produce flux in the core. If we choose to use a former with a high permeability, the reluctance will be low, and so it will become easier to produce flux in the core. Thus, we can regard the reluctance as the resistance to the flow of flux.

To return to Equation (3.37), the term on the left-hand side is the number of turns times the current flowing through them. If we continue drawing our parallel with electrostatics, the ampere-turns appear to be the equivalent of potential. We defined potential as the work done in moving a unit charge through an electric field. So, are the ampere-turns equal to the work done in moving a unit pole through a magnetic field?

As we saw at the start of this chapter, the force on a unit pole placed in a magnetic field is (Equation (3.3))

$$F = p_2 H$$

Now, the magnetic field in the core is (Equation (3.35))

$$H = \frac{NI}{l} \text{ A m}^{-1}$$

or

$$H = \frac{NI}{2\pi r} \text{ A m}^{-1} \tag{3.39}$$

where r is the average radius of the former. So, if we move a unit north pole around the former, we have to do work of

$$\text{Work done} = \text{force} \times \text{distance}$$

$$= 1 \times \frac{NI}{2\pi r} \times 2\pi r$$

$$= NI \text{ At}$$

Thus, the magnetic potential (the work done) is given by

$$V_m = NI \text{ At} \tag{3.40}$$

We can now rewrite Equation (3.37) as

$$V_m = \phi \times S \tag{3.41}$$

So, the magnetic potential is equal to the product of the magnetic flux and reluctance. This is very similar to Ohm's law which applies to electroconductive fields. We will put this similarity to good use when we consider transformers in Chapter 7. (Some texts refer to the magnetic potential as the magneto-motive force or mmf. However, this gives an image of the mmf forcing the flux around the magnetic circuit. As we have seen, the mmf is not a force, but rather is a measure of the energy required to move a pole around a magnetic field. As the term mmf can be a source of confusion, we will refer to it as magnetic potential.)

Example 3.5

A toroidal coil is wound on a ferrite former with inner radius 4 cm and outer radius 6 cm. The coil has 2000 turns, and the core has a relative permeability of 200. Determine the current required to produce a mean flux of 1 mWb in the core.

Solution

The magnetic potential is the product of the flux and the reluctance of the core. Now, before we can find the reluctance, we need to find the average circumference and cross-sectional area of the core. The average radius of the core is

$$\text{Average radius} = \frac{4+6}{2} \text{ cm}$$

$$= 5 \text{ cm}$$

Thus, the average length of the magnetic path is

$$l = 2\pi \times 5 \times 10^{-2}$$

$$= 0.314 \text{ m}$$

The former has a circular cross-sectional area of radius

$$r = \frac{6-4}{2} \text{ cm}$$

$$= 1 \text{ cm}$$

and so the reluctance of the core is (Equation (3.38))

$$S = \frac{1}{\mu \times \text{area}}$$

$$= \frac{1}{200 \times 4\pi \times 10^{-7} \times \pi \times \left(1 \times 10^{-2}\right)^2}$$

$$= 4 \times 10^6 \text{ At Wb}^{-1}$$

Thus, the ampere-turns required are (Equation (3.41))

$$V_m = 1 \times 10^{-3} \times 4 \times 10^6$$

$$= 4 \times 10^3 \text{ At}$$

As the coil has 2000 turns, the coil current is

$$I = \frac{4 \times 10^3}{2000}$$

$$= 2 \text{ A}$$

Let us take a moment to examine these figures. A coil current of 2 A through 2000 turns gives a flux density of only 1 mWb. Clearly a large current and lots of wire are required to generate high magnetic fields.

3.11   INDUCTANCE

So far we have only considered coils that have a steady d.c. current passing through them. This introduced us to the idea of magnetic flux, the magnetic flux density and magnetic field strength. Although d.c. circuits sometimes use coils, we more usually find them in a.c. circuits. In such circuits, we tend to characterize coils by a term called inductance.

When a d.c. voltage energizes a coil, a current flows which sets up a magnetic field around the coil. This field will not appear instantaneously as it takes a certain amount of time to produce the field. After the initial transient has passed, the resistance of the wire that makes up the coil will limit its current.

Let us now consider a very low-resistance coil connected to a source of alternating voltage. As the coil resistance is very low, the coil should appear to be a short-circuit. This should result in a lot of current flowing! However, what we find is that the current taken by the coil depends on the frequency of the source – high frequencies result in low currents. Thus, some unknown property of the coil restricts the current.

In 1831, a British physicist, chemist and great experimenter called Michael Faraday (1791–1867) was investigating electromagnetism. As a result of his experiments, Faraday proposed that a changing magnetic field induces an emf into a coil. This was one of the most significant discoveries in electrical engineering, and it is the basic principle behind transformers and electrical machines. (Faraday's achievement is even more remarkable in that all of his work resulted from experimentation, and not mathematical derivation.)

Faraday's law formalizes this result as

$$e \propto \frac{d}{dt}(N\phi) \qquad (3.42)$$

where N is the number of turns in the coil and $N\phi$ is known as the flux linkage. So, the induced emf depends on the rate of change of flux linkages, i.e., the higher the frequency, the higher the rate of change, the larger the induced emf. As this emf serves to oppose the voltage that produces it, Equation (3.42) is often modified to

$$e = -\frac{d}{dt}(N\phi) \qquad (3.43)$$

So, if we have a coil connected to a source of alternating voltage, the coil takes current which produces a magnetic field. This magnetic field produces an alternating magnetic flux, by virtue of $B = \mu H$, so generating a back-emf in the coil (note the minus sign in Equation (3.43)). This back-emf opposes the voltage producing it, and this reduces the current taken from the supply.

An alternative 'circuits' expression for the back-emf is

$$e = -\frac{d}{dt}(Li) \qquad (3.44)$$

where L is the inductance of the coil and i is the alternating current taken by the coil. Now, by equating these two expressions, we have

$$-\frac{d}{dt}(N\phi) = -\frac{d}{dt}(Li)$$

and so,

$$L = N \frac{d\phi}{di} \qquad (3.45)$$

Thus, the inductance is the flux linkage per unit current. The unit of inductance is the Henry, named after the American physicist Joseph Henry (1791–1878) who invented the electromagnetic telegraph as well as extensively studying electromagnetic phenomena.

We will now go on to examine the inductance of a simple coil, which may be air-cored or iron-cored. However, what is seldom appreciated is that a piece of wire can have an inductance – termed self-inductance. Then, we will also examine the inductance of some transmission lines.

3.11.1 Simple Coil

A simple coil consists of several turns on wire wound around a former. As we have just seen, the inductance is defined as the flux linkage per unit current, i.e.,

$$L = N \frac{d\phi}{di}$$

where N is the number of turns in the coil. When we considered solenoids, we saw that the flux density varies along the axis of the coil. However, if the coil is very long, the field at the centre of the coil is

$$H = \frac{NI}{l}$$

and so,

$$B = \mu \frac{NI}{l}$$

As B is the flux density, i.e., $B = \phi/A$, we can write

$$L = N \frac{dB}{di} A$$

$$= NA \, \mu \, \frac{N}{l} \frac{di}{di}$$

$$= \frac{N^2 \mu A}{l} \tag{3.46}$$

where A is the cross-sectional area of the coil. Although Equation (3.46) gives the inductance of a long coil, this equation is an approximation. This is because it assumes that the field is constant throughout the coil, and it neglects the effects of flux leakage.

Example 3.6

An air-cored coil has a diameter of 2 cm and a length of 10 cm. The number of turns per cm is 50. Estimate the inductance of the coil. If a powdered ferrite core is added, with μ_r of 100, estimate the new value of inductance. Determine the new coil length if the ferrite-core coil is to have the same value of inductance as the original air-cored coil.

Solution

The coil is 10 cm long and has 50 turns per cm. So, the total number of turns on the coil is 500. From Equation (3.46), the inductance is

$$L = \frac{500^2 \times 4\pi \times 10^{-7} \times \pi \times \left(1 \times 10^{-2}\right)^2}{10 \times 10^{-2}}$$

$$= 1\,\text{mH}$$

If we now insert a ferrite core the value of μ will increase. As the core has $\mu_r = 100$, the new inductance is

$$L = 1 \times 10^{-3} \times 100$$

$$= 100\,\text{mH}$$

We now require the iron-cored coil to have the same inductance as the original air-cored coil. To do this, we will retain the same number of turns per metre (5×10^3) and reduce the length of the coil. So,

$$L = \frac{\left(5 \times 10^3 \times 1\right)^2 \times \mu_0 \mu_r \times A}{l} = 1\,\text{mH}$$

After some rearranging, this gives

$$l = \frac{1 \times 10^{-3}}{(5 \times 10^3)^2 \times \mu_0 \mu_r \times A}$$

$$= 1 \times 10^{-3}\,\text{m}$$

$$= 1\,\text{mm}$$

Thus, the number of turns is only 5, and the length of the coil is 1 mm. This means that the coil length is much smaller than the diameter, and so we must treat the calculated inductance value with suspicion. However, this example does show the effects of a ferrite core on the length of a coil.

3.11.2   SELF-INDUCTANCE OF A SINGLE WIRE

We have just seen that inductance is equal to the flux linkages per unit current. The example at the end of Section 3.6 examined the variation of magnetic field both inside and outside the wire. As we saw, the magnetic field inside the wire is associated with a fraction of the current flowing through the wire. So, as we have different fields inside and outside the wire, we should expect that there are two components to the inductance of a piece of wire,

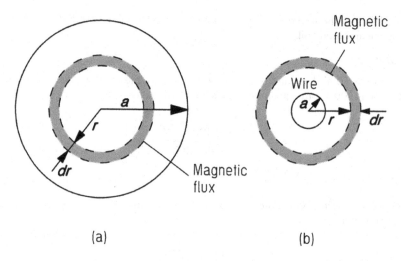

FIGURE 3.14 (a) Cross section of a current-carrying wire and (b) magnetic flux outside a current-carrying wire.

Let us initially examine the field inside the wire. Figure 3.14a shows a cross section of the wire, in which we have a circular path of radius r and thickness dr. If we assume a constant current density, the current enclosed by this loop is

$$I' = \frac{I}{\pi a^2} \pi r^2$$

This current will produce a magnetic field of

$$H' = \frac{I}{\pi a^2} \pi r^2 \frac{1}{2\pi r}$$

$$= \frac{Ir}{2\pi a^2}$$

which gives a flux density of

$$B' = \frac{\mu_0 Ir}{2\pi a^2} \text{ at a radius of } r$$

We require an expression linking the flux, at radius r, to the current producing the flux. So, in order to find the flux, we will consider and incremental ring of thickness dr and radius r as shown in Figure 3.14a. The area that the flux sees is $dr \times$ length, and so the flux through the ring is

$$d\phi = B' \times dr \times \text{length}$$

$$= \frac{\mu_0 Ir}{2\pi a^2} dr \times \text{length}$$

This flux is generated by the fraction of the total current flowing through the wire. Now, in order to find the fractional flux linkage, we have to know how much of the wire is linked with this flux. Thus, we need to multiply the incremental flux by a fraction. Specifically, the fractional flux linkage, $d\ddot{U}$, is

$$d\ddot{U} = d\phi \times \frac{\pi r^2}{\pi a^2}$$

$$= \frac{\mu_0 I r}{2\pi a^2} dr \times \text{length} \times \frac{\pi r^2}{\pi a^2}$$

$$= \frac{\mu_0 I}{2\pi a^4} r^3 \times dr \times \text{length}$$

So, the total flux linkage inside the wire is

$$\ddot{U} = \frac{\mu_0 I}{2\pi a^4} \int_0^a dr \times \text{length}$$

$$= \frac{\mu_0 I}{2\pi a^2} \left(\frac{r^4}{4} \right)_0^a \times \text{length}$$

$$= \frac{\mu_0 I}{8\pi a^4} \frac{a^4}{4} \times \text{length}$$

$$= \frac{\mu_0 I}{8\pi} \times \text{length}$$

Hence, the total internal inductance per unit length is

$$L' = \frac{\mu_0}{8\pi} \text{ H m}^{-1} \tag{3.47}$$

This is the internal inductance of the wire. As this equation shows, it is independent of the diameter of the wire.

Let us now turn our attention to the flux outside the wire. Application of Ampere's law gives the field strength at a radius r as

$$H' = \frac{I}{2\pi r}$$

and so the flux density at this radius is

$$B' = \frac{\mu_0 I}{2\pi r}$$

Let us again consider an incremental tube of radius r and thickness dr – see Figure 3.14b. The fractional flux through the surface of this ring is

$$d\phi = B' \times dr \times \text{length}$$

$$= \frac{\mu_0 I}{2\pi r} dr \times \text{length}$$

This is the fractional flux linkage generated by the total current flowing in the wire. So, the fractional inductance per unit length is

$$dL' = \frac{d\phi}{I}$$

$$= \frac{\mu_0}{2\pi} \frac{dr}{r}$$

and the total external inductance per unit length is

$$L' = \frac{\mu_0}{2\pi} \int\limits_{a}^{\infty} \frac{dr}{r}$$

$$= \frac{\mu_0}{2\pi} \left. \ln r \right|_{a}^{\infty} = \frac{\pi_0}{2\pi} \left(\ln \infty - \ln a \right)$$

$$= \text{infinity} \qquad\qquad (3.48)$$

This rather surprising result arises because we assumed that the field is zero at infinity. (A similar situation arose when we examined line charges in Section 2.6.) However, we should remember that there must be a return path for the current – it may be a very long way away, but it must be there. What usually happens is that the wire has a ground-plane at a certain distance from the conductor. We will meet this situation again when we consider the inductance of microstrip later in this section.

Example 3.7

Determine the inductance per unit length of 5 mm diameter copper wire.

Solution

We have wire of diameter 5 mm, and so the radius is 2.5 mm. The internal self-inductance of the wire is independent of radius, and so

$$L_{int} = \frac{\mu_0}{8\pi}$$

$$= \frac{4\pi \times 10^{-7}}{8\pi}$$

$$= 50 \text{ nH}$$

As regards the external inductance, we can write

$$L_{ext} = \frac{\mu_0}{2\pi} (\ln d - \ln a)$$

Let us calculate the total inductance assuming the magnetic field is zero at certain distances.

Distance	External Inductance (nH)	Total Inductance (nH)	Percentage Increase
$2a$	140	190	-
$20a$	600	650	242
$200a$	1060	1110	71
$2000a$	1520	1570	41
$2 \times 10^4 a$	2000	2050	30
$2 \times 10^5 a$	2440	2490	31

This table shows that the percentage increase in total inductance falls as the distance from the wire increases. We should also note that the internal inductance become insignificant when compared with the external inductance.

3.11.3 COAXIAL CABLE

Figure 3.15 shows a cross section through a length of coaxial cable. Now, current in the inner conductor generates a magnetic field in the inner conductor, and in the dielectric between the inner and outer conductors. The outer conductor is usually earthed and effectively shields the signal on the inner conductor from any external interference. Thus, the field at the outer conductor is zero.

As we have seen in the previous section, there will be two parts to the inductance; the inductance of the inner conductor and the inductance due to the magnetic field in the dielectric. As regards the inductance of the inner conductor, the last section gave the value of

$$L_{int} = \frac{\mu_0}{8\pi} \text{ H m}^{-1}$$

To find the external inductance, we will follow a similar procedure to that used in the last section. Thus, the flux density at a radius r in the dielectric is

$$B = \mu_0 \frac{I}{2\pi r} \qquad (3.49)$$

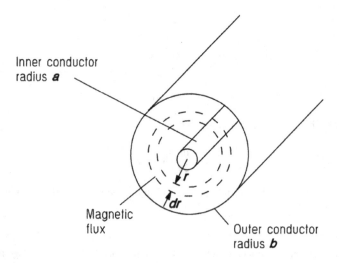

Inner conductor
radius **a**

Magnetic
flux

Outer conductor
radius **b**

FIGURE 3.15 Cross section through a length of coaxial cable.

With this flux density, the flux through a small incremental ring of radius r and thickness dr is

$$d\phi = B' \times dr \times \text{length}$$

$$= \frac{\mu_0 I r}{2\pi a^2} dr \times \text{length}$$

This is the fractional flux linkage generated by the current I flowing in the inner conductor. Thus, the fractional external inductance per unit length is

$$dL' = \frac{d\phi}{I}$$

$$= \frac{\pi_0}{2\pi} \frac{dr}{r}$$

and so the total external inductance per unit length is

$$L_{\text{ext}} = \frac{\mu_0}{2\pi} \int_a^b \frac{dr}{r}$$

$$= \frac{\mu_0}{2\pi} \left. \ln r \right|_a^b$$

$$= \frac{\mu_0}{2\pi} (\ln b - \ln a)$$

$$= \frac{\mu_0}{2\pi} \ln (b/a) \qquad (3.50)$$

Thus, the total inductance per unit length is

$$L' = \frac{\mu_0}{8\pi} + \frac{\mu_0}{2\pi} \ln(b/a) \qquad (3.51)$$

We should note that the internal inductance can be insignificant when compared with the external inductance. Also, at high frequencies the current tends to crowd towards the surface of the inner conductor. This is known as the skin effect, and when it occurs, the internal inductance tends to be zero. Thus, the inductance per unit length approximates to

$$L' = \frac{\mu_0}{2\pi} \ln(b/a) \text{ H m}^{-1} \qquad (3.52)$$

As we saw in Section 2.9, such a cable also has a capacitance given by (Equation (2.37))

$$C' = \frac{2\pi\varepsilon_0\varepsilon_r}{\ln(b/a)} \text{ F m}^{-1}$$

Now, if we multiply these two equations together, we get

$$L'C' = \frac{\mu_0}{2\pi} \ln(b/a) \frac{2\pi\varepsilon_0\varepsilon_r}{\ln(b/a)}$$

$$= \mu_0\varepsilon_0 \qquad (3.53)$$

So, the product of inductance and capacitance results in the product of permeability and permittivity. This is an interesting result, which we will return to in the next section.

Example 3.8

A 500 m length of coaxial cable has an inner conductor of radius 2 mm, and an outer conductor of radius 1 cm. A non-ferrous dielectric separates the two conductors. Determine the inductance of the cable.

Solution

The dielectric is non-ferrous, and so the permeability of the dielectric is that of free-space. So, the inductance per unit length is (Equation (3.51))

$$L' = \frac{\mu_0}{8\pi} + \frac{\mu_0}{2\pi} \ln(b/a)$$

$$= 50 \times 10^{-9} + 200 \times 10^{-9} \ln 5$$

$$= 50 \times 10^{-9} + 322 \times 10^{-9} \ln 5$$

$$= 372 \text{ nH m}^{-1}$$

As the cable is 500 m long, the inductance is

$$L = 372 \times 500 \text{ nH}$$

$$= 186 \text{ }\mu\text{H}$$

At frequencies above about 1 MHz, we can neglect the internal inductance. Thus, the inductance of the cable becomes

$$L = 161 \text{ }\mu\text{H}$$

The dimensions of this cable are the same as the cable in Example 2.9 used in Section 2.9 in the last chapter.

3.11.4 TWIN FEEDER

Figure 3.16 shows a section through a length of twin feeder. The left-hand conductor carries a current of I A, while the return conductor (the right-hand one) carries a current of $-I$ A. Both conductors will produce magnetic fields in the space between the wires, and so there will be two components to the flux density at any particular point.

As before, there will be two components to the inductance; the internal inductance of each conductor and the external inductance of each conductor. We have already calculated the internal inductance of a conductor. However, to find the external inductance, we need to generate an equation linking the total flux with the current in one of the conductors.

To analyze the situation, we will use the principle of superposition to find the total flux density at some point between the two wires. So, the flux density at a radius r due to the current in the left-hand conductor is

$$B' = \frac{\mu_0 I}{2\pi r} \tag{3.54}$$

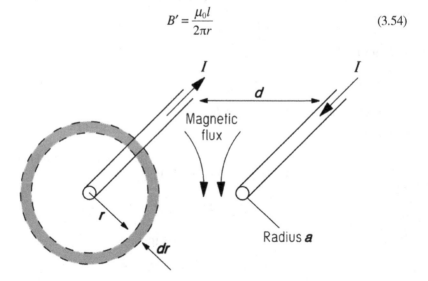

FIGURE 3.16 Cross section through a length of twin feeder.

The right-hand conductor generates a flux density at the same point of

$$B = \frac{\mu_0 l}{2\pi(d-r)} \tag{3.55}$$

acting in the same direction. So, the total field at a radius of r from the left-hand conductor is

$$B = \frac{\mu_0 l}{2\pi r} + \frac{\mu_0 l}{2\pi(d-r)}$$

Thus, the fractional flux through an incremental ring of radius r and thickness dr is

$$d\phi = \left(\frac{\mu_0 l}{2\pi r} + \frac{\mu_0 l}{2\pi(d-r)} \right) dr \times \text{length}$$

and so the fractional inductance per unit length is

$$dL' = \left(+ \frac{\mu_0 l}{2\pi(d-r)} \right) dr \tag{3.56}$$

In order to find the total inductance per unit length, we need to integrate this equation between the limits a and $d-a$. So,

$$L' = \frac{\mu_0}{2\pi} \int_{a}^{d-a} \left(\frac{I}{r} + \frac{I}{d-r} \right) dr$$

$$= \frac{\mu_0}{2\pi} \left. \ln r - \ln(d-r) \right|_{a}^{d-a}$$

$$= \frac{\mu_0}{2\pi} \left(\ln(d-a) + \ln a + \ln a + \ln(d-a) \right)$$

$$= \frac{\mu_0}{\pi} \ln\left(\frac{d-a}{a} \right) \mathrm{H\,m}^{-1} \tag{3.57}$$

As we have two conductors, the internal inductance, from our previous studies, is doubled. Thus,

$$L' = \frac{\mu_0}{4\pi} \mathrm{H\,m}^{-1}$$

And so the total inductance per unit length of twin feeder is

$$L' = \frac{\mu_0}{4\pi} + \frac{\mu_0}{\pi} \ln\left(\frac{d-a}{a} \right) \mathrm{H\,m}^{-1} \tag{3.58}$$

If we can ignore the internal inductance, we get

$$L' = \frac{\mu_0}{\pi} \ln\left(\frac{d-a}{a}\right) \mathrm{H\,m^{-1}} \tag{3.59}$$

As we saw in Section 2.9, this type of cable also has capacitance given by (Equation (2.44))

$$C' = \frac{\pi\varepsilon_0}{\ln\left(\dfrac{d-a}{a}\right)} \mathrm{F\,m^{-1}}$$

If we multiply these two equations together, we get

$$L'C' = \frac{\mu_0}{\pi} \ln\left(\frac{d-a}{a}\right) \frac{\pi\varepsilon_0}{\ln\left(\dfrac{d-a}{a}\right)}$$

$$= \mu_0\varepsilon_0$$

This is the same result as that obtained with coaxial cable, Equation (3.53). This has the making of a rule; in a transmission line, the product of capacitance and inductance equals the product of the permittivity and permeability, i.e.,

$$L'C' = \mu\varepsilon$$

As the speed of light is $1/\sqrt{\mu\varepsilon}$, we can write

$$c = \frac{1}{\sqrt{L'C'}} = \frac{1}{\sqrt{\mu\varepsilon}} \tag{3.60}$$

This shows that signals propagate down the line at the speed of light in the dielectric (slower than they travel in free-space). We will use this rule when we consider microstrip in the next section.

Example 3.9

A 200 m length of feeder consists of two 2-mm radius conductors separated by a distance of 20 cm. Determine the inductance of the arrangement.

Solution

If we assume that the conductors are in air, the permeability of the dielectric surrounding them is that of free-space. Thus, from Equation (3.58), we get

$$L' = 100 \times 10^{-9} + 400 \times 10^{-9} \ln 99$$

$$= 100 \times 10^{-9} + 400 \times 10^{-9} \times 4.6$$

$$= 100 \times 10^{-9} + 1840 \times 10^{-9}$$

$$= 1.94 \,\mu\text{H}\,\text{m}^{-1}$$

So, the total inductance of the cable is

$$L = 1.94 \times 10^{-6} \times 200$$

$$= 388 \,\mu\text{H}$$

At high frequencies, the internal inductance tends to zero, and so the cable inductance tends to

$L = 368 \,\mu\text{H m}^{-1}$

The dimensions of this cable are identical to that used in Example 2.10 in Section 2.9. So, twin feeder has inductance and capacitance.

3.11.5 MICROSTRIP LINES

Let us now consider microstrip. This was first introduced in Section 2.9 when we considered the capacitance of a printed circuit track over a ground-plane. To find the inductance of this arrangement, we would have to plot the magnetic field surrounding the track. This involves a considerable amount of work, which we can avoid by using the rule introduced at the end of the last section.

When we considered coaxial cable and twin feeder, we found that $L'C' = \mu\varepsilon$. So, if we can determine the capacitance, we can find the inductance per unit length. In Section 2.9, we found that we could approximate the capacitance to that of a parallel plate capacitor (Equation (2.48))

$$C' = \frac{\varepsilon_0 \varepsilon_r w}{h} \,\text{F}\,\text{m}^{-1}$$

Or, if the track width is much less than the thickness of the board, the capacitance of a cylindrical wire over a ground-plane (Equation (2.49))

$$C' = \frac{2\pi\varepsilon_0 \varepsilon_r}{\ln\left(\dfrac{h}{w}\right)} \,\text{F}\,\text{m}^{-1}$$

Thus, the inductance lies between

$$L' = \frac{\mu_0 h}{w} \,\text{H}\,\text{m}^{-1} \tag{3.61}$$

for the parallel plate approximation, and

$$L' = \frac{\mu_0}{2\pi} \ln\left(\frac{h}{w}\right) H m^{-1} \tag{3.62}$$

for the wire above ground approximation.

Example 3.10

A 3-mm wide track is etched on one side of some double-sided printed circuit board. The thickness of the board is 2 mm, and the dielectric has a relative permeability of 1. Determine the inductance per cm.

Solution

As the track width is of the same order of magnitude as the board thickness, we must use Equation (3.61) to give

$$L' = \frac{\mu_0 h}{w} H m^{-1}$$

$$= 4\pi \times 10^{-7} \frac{2 \times 10^{-3}}{3 \times 10^{-3}}$$

$$= 8.4 \times 10^{-7} H m^{-1}$$

$$= 8.4 \ nH \ cm^{-1}$$

This neglects the effect of the internal inductances given by

$$L' = \frac{\mu_0}{8\pi}$$

$$= 50 \ nH \ m^{-1}$$

$$= 0.5 \ nH \ cm^{-1}$$

Thus, the total inductance is-

$$L' = 8.9 \ nH \ cm^{-1}$$

3.11.6 ENERGY STORAGE

In the same way that energy is stored in an electric field, energy can also be stored in a magnetic field. This is useful in switch-mode power supplies. To find the stored energy, let us take an inductor connected to a d.c. source. This inductor will take a certain amount of current, limited by the resistance of the coil. If we increase the current by a small amount dI in time dt, the flux causes a back-emf given by

$$dV = L \frac{dI}{dt} \tag{3.63}$$

As the current flowing through the coil is I, the instantaneous power supplied is

$$I \, dV = LI \, \frac{dI}{dt} \qquad (3.64)$$

This power is supplied in time dt, and so the energy supplied in raising the current from I to $I + dI$ is

$$dE = I \, dV \, dt$$

$$= L \, I \frac{dI}{dt} dt$$

$$= L \, I \, dI \qquad (3.65)$$

Thus, we can find the energy supplied in raising the current from zero to I by integrating Equation (3.65). So,

$$\text{stored energy} = \int_0^1 L \, I \, dI$$

$$= \frac{1}{2} \, LI^2 \text{ J} \qquad (3.66)$$

It is interesting to compare this equation with that obtained for the energy stored in a capacitor, Equation (2.51), stored energy $= \frac{1}{2} \, CV^2$ joule.

Apart from the obvious difference that inductance replaces capacitance and current replaces voltage, the two expressions are similar. As we have already seen, a current produces a magnetic field, and coils certainly have magnetic fields. However, a voltage produces an electric field and capacitors have electric fields. To take this one stage further, the stored energy per unit volume in a capacitor is given by energy $= \frac{1}{2} \, DE$ J m^{-3} and so we can postulate that the stored energy per unit volume in a coil is

$$\text{energy} = \frac{1}{2} BH \, \text{J} \, \text{m}^{-3} \qquad (3.67)$$

In order to prove this, we can substitute for inductance from Equation (3.46) into Equation (3.66) to give energy $= \frac{1}{2} \, LI^2$

$$= \frac{1}{2} \frac{N^2 \mu A}{l} I^2$$

$$= \frac{1}{2} \frac{N^2 I^2}{l^2} \mu \text{ area} \times \text{length}$$

$$= \frac{1}{2} H^2 \mu \text{ area} \times \text{length}$$

$$= \frac{1}{2} BH \, \text{J} \, \text{m}^{-3}$$

So, our supposition was correct. This is an important result in that it shows a duality between electrostatic and electromagnetic fields.

Example 3.11

A 10 mH inductor takes a current of 1.5 A. Determine the energy stored in the magnetic field.

Solution

As the inductance is quoted, we can use Equation (3.66) to give

$$\text{energy} = 1/2\,LI^2$$

$$= 1/2 \times 10 \times 10^{-3} \times 1.5^2$$

$$= 11.25\,\text{mJ}$$

There is one major difference between capacitors and inductors: the energy stored in a capacitor stays there for a very long time (it eventually decays away to zero due to leakage), whereas the energy stored in an inductor disappears almost as soon as we remove the current. This is why capacitors are preferred as energy storage devices. (It is true that computer memory boards and uninterruptable power supplies use capacitors to store energy. However, they can only store enough for very short duration use. If we want a continuous supply, we must use chemical batteries such as lead-acid or lithium cells.)

3.11.7   FORCE BETWEEN TWO MAGNETIC SURFACES

In the last section, we determined the energy stored in the magnetic field of a coil. We were able to draw a comparison with the energy stored in a capacitor to predict the energy density in a magnetic field. Now that we are considering the force between magnetic surfaces, can we use the same procedure?

When we considered electrostatic force, we found that the force between two charged plates is given by (Equation (2.55))

$$F = \tfrac{1}{2}QE\ \text{N}$$

So, we can postulate that the force between two magnetic surfaces is given by

$$F = \tfrac{1}{2}\phi H\ \text{N}$$

The method we will use is the same as the one we used to find the electrostatic force. So, let us consider the arrangement shown in Figure 3.17. This stores a certain amount of energy given by (Equation (3.67))

$$\text{energy} = \tfrac{1}{2}BH\ \text{J}\,\text{m}^{-3}$$

or

$$\text{energy} = \tfrac{1}{2}BH \times \text{area}\,l$$

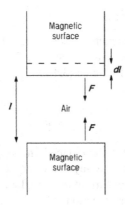

FIGURE 3.17 Force between magnetic surfaces.

Now, there will be a force of attraction between the two surfaces. If we move the top surface by a small amount dl, we do work against the attractive force. This work done must equal the change in stored energy. Thus,

$$F\mathrm{d}l = \tfrac{1}{2} BH \times \text{area} \times (l + \mathrm{d}l) - \tfrac{1}{2} BH \times \text{area} \times l$$

$$= \tfrac{1}{2} BH \times \text{area} \times \mathrm{d}l$$

Hence,

$$F = \tfrac{1}{2} BH \times \text{area}$$

$$= \tfrac{1}{2}\phi H \text{ N} \tag{3.68}$$

which agrees with our earlier prediction. The flux and field strength in Equation (3.68) relate to the air gap between the two surfaces. So we can write

$$\text{force} = \frac{1}{2}\frac{B^2}{\mu_0} \times \text{area} \tag{3.69}$$

Example 3.12

The flux density in an air gap between two magnetic surfaces is 5 mWb m^{-2}. Determine the force between the two surfaces if their cross-sectional area is 5 cm?

Solution

We can use Equation (3.69) to give

$$\text{force} = \frac{1}{2}\frac{B^2}{\mu_0} \times \text{area}$$

$$= \frac{1}{2}\frac{\left(5\times10^{-3}\right)^2}{4\pi\times10^{-7}} \times 5\times10^{-4}$$

$$= 5\times10^{-3} \text{ N}$$

3.11.8   LOW-FREQUENCY EFFECTS

At the start of Section 3.10, inductance was introduced as a parameter that limits the current when a coil is connected to an a.c. supply. We saw that a back-emf limits the current with the emf given by (Equation (3.44))

$$e = -L\frac{di}{dt}$$

where the minus sign shows that the induced emf opposes the supply voltage. An alternative way of looking at this is to say that when a coil is connected to an alternating voltage source, an alternating current flows in the coil. We can find the current by solving the following differential equation

$$v_s(t) = L\frac{di}{dt} \tag{3.70}$$

(We should note that the minus sign is missing because Equation (3.70) uses the supply voltage.)

Figure 3.18a shows an inductor connected to an alternating supply. As the source is varying with time, we can write

$$i(t) = I_{pk}\sin\omega t$$

and so Equation (3.70) becomes

$$v_s(t) = L\,I_{pk}\,\omega\,\cos\omega t$$
$$= L\,I_{pk}\,\omega\,\sin(\omega t + 90°) \tag{3.71}$$

So, when connected to an alternating source, the inductor allows a current to flow with the supply voltage leading the current by 90°. (Figure 3.18b shows the relationship between the supply voltage and the inductor current.)

We should note that the back-emf is directly proportional to the angular frequency of the source. We can intuitively reason that the coil current depends on the magnitude of this back-emf: if the frequency is low, the back-emf is small and so the current taken will be high; if the frequency is high, the back-emf is large and so the current taken will be small. We can formalize this by defining the reactance of the inductor, X_L, as

$$X_L = \omega L$$
$$= 2\pi f L \tag{3.72}$$

By combining this result with Equation (3.71) we get, after some rearranging,

$$v_s(t) = i(t)\,X_L\underline{/90°}$$

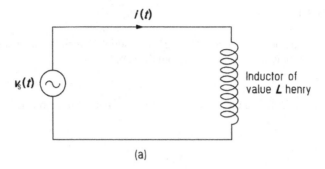

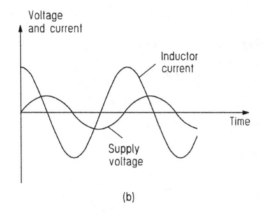

FIGURE 3.18 (a) An inductor connected to an a.c. supply and (b) relationship between inductor current and voltage.

and so,

$$i(t) = \frac{v_s(t)}{X_L} \underline{/90°} \tag{3.73}$$

So, when an inductor is connected to an a.c. source, it provides a high-resistance path for a.c. signals. If there is also a d.c. voltage, this will allow current to flow. So, we can use an inductor to block a.c. signals but allow d.c. to pass through.

This blocking ability can be used in d.c. power supplies. In such power supplies, a d.c. voltage is produced from a rectified a.c. signal. To remove any a.c. content, we can place an inductor in series with the d.c. output. (The inductor acts as a high resistance to the a.c., but lets the d.c. through without any effect.)

The only problem with this arrangement is that all the current taken from the supply has to pass through the inductor. This current can be quite high (greater than 10 A in some instances) and so the inductor must use heavy-duty wire. This is why power supplies use smoothing capacitors connected between the live and ground terminals.

It is quite common to find coils in the supply rail of high-frequency circuits. These coils, known as radio-frequency chokes, act to stop any signals from getting to the power rails where they might cause trouble to other circuits.

Example 3.13

A 50 mH inductor is connected to a 12 V a.c. supply which has a frequency of 100 Hz. Determine the current taken from the supply.

Solution

The frequency of the supply is 100 Hz, and so the angular frequency is

$$\omega = 2\pi f$$

$$= 2\pi \times 100$$

$$= 200\pi \text{ rad s}^{-1}$$

Now, the reactance of the inductor is given by

$$X_L = \omega L$$

$$= 200\pi \times 50 \times 10^{-3}$$

$$= 31.4 \ \Omega$$

Thus, the supply current is

$$i_s = \frac{V_s}{X_L}$$

$$= \frac{12}{31.4}$$

$$= 382 \text{ mA}$$

3.12   SOME APPLICATIONS

Figure 3.19 shows a cut-away of the Joint European Torus, or JET, tokamak fusion reactor at Culham, Oxfordshire. This is an experimental machine in which large amounts of energy can be liberated by the fusion of the hydrogen isotopes deuterium and tritium. (This is the same process that goes on in the Sun.) To initiate fusion, it is necessary to control a plasma with a temperature in excess of 200 million°C! As the plasma consists of positively charged nuclei, an obvious way of confining it is to use a magnetic field (Figure 3.20)

As can be seen in the figure, the reactor vessel is toroidal in shape with an inner radius of 1.25 m and an outer radius of 3 m. This vessel is surrounded by 32 electro-magnetic coils arranged around the outside. Each coil consists of 24 turns of wire carrying a current of 67 kA, which results in a maximum field of 3.4 Wb m^{-2} at the centre of the vessel. Eight large iron-cored transformers generate an additional field that serves to confine, as well as heat, the plasma. (In effect, the plasma forms a single-turn secondary, with energy being supplied from the primary coil to the plasma via the iron limbs.) This arrangement has resulted in plasma currents of up to 7 MA, and this serves to heat the plasma to about 50 million°C. Further heating

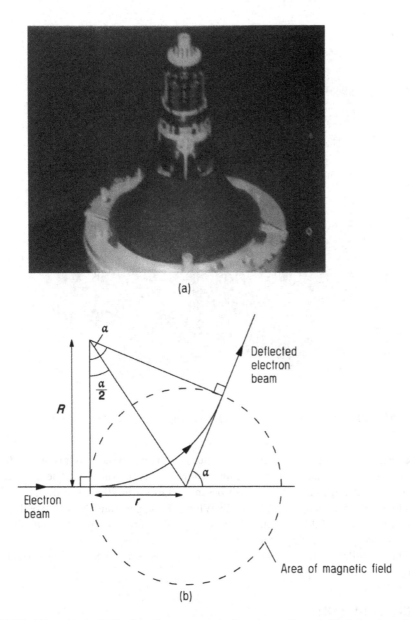

(a)

(b)

FIGURE 3.19 (a) A CRT with electromagnetic focusing coils and (b) deflection of an electron beam by a magnetic field.

is provided by 20 MW of neutral particle injection (approximately 140 keV) and 24 MW of radio frequency power (23–57 MHz).

Although JET is not designed to break even, i.e., generate more energy than it consumes, the project has produced some fusion energy. The conclusion is that a larger version of the machine could generate useful power. In view of this, it is proposed

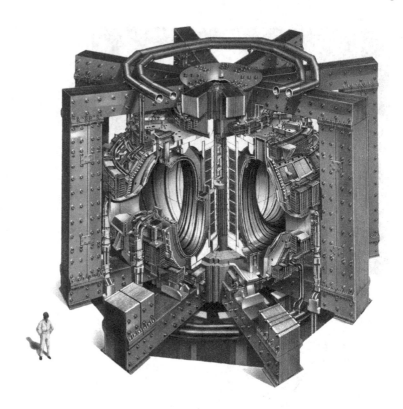

FIGURE 3.20   Schematic cutaway of the Joint European Torus. (© EUROfusion.)

to construct a new torus with an inner radius of 3 m and an outer radius of 8 m. If all goes according to plan, the new machine, known as the International Thermonuclear Experimental Reactor, or ITER, will begin operations in 2025. The design of this machine calls for a magnetic field of 13 Wb m^{-2} to be generated by superconducting coils carrying 40–60 kA of current, resulting in a stored energy of about 100 GJ! This is clearly engineering on a grand scale.

On a smaller scale, it is possible to deflect an electron beam using permanent magnets or electromagnets.

3.13   SUMMARY

We started this chapter by examining Coulomb's law as applied to magnetostatics. This introduced us to the general ideas of magnetic flux, magnetic flux density and magnetic field strength. This initial study was only concerned with isolated magnetic monopoles.

We then went on to consider the magnetic field produced by a current carrying wire – electromagnetism. This introduced us to the Biot-Savart law which is a fundamental law of electromagnetism. We then went on to review our ideas of magnetic flux density and magnetic field strength. The relevant formulae are summarized here:

$$dH = \frac{Idl}{4\pi r^2}\sin\theta \tag{3.76}$$

$$dB = \mu\frac{Idl}{4\pi r^2}\sin\theta\, p_N \tag{3.77}$$

$$dF = pdB \tag{3.78}$$

We then examined the magnetic field produced by a variety of current-carrying wires: a long straight conductor; a circular conductor; and a solenoid, or coil. As well as introducing us to applications of the Biot-Savart law, these examples also introduced us to Ampère's law. We then met magnetic potential. This was defined as the work done in moving a unit pole around a magnetic field, and introduced us to the idea of reluctance as the resistance to the flow flux.

In Section 3.11, we were introduced to the idea of inductance. We determined the inductance of various arrangements: simple coil; an isolated wire; coaxial cable; twin feeder and microstrip lines. The inductances are reproduced here:

Simple coil

$$L = \frac{N^2\mu A}{l}H \tag{3.79}$$

Isolated wire

$$L' = \frac{\mu_0}{8\pi}\,\mathrm{H\,m^{-1}}\,(\text{internal}) \tag{3.80}$$

Coaxial cable

$$L' = \frac{\mu_0}{8\pi} + \frac{\mu_0}{2\pi}\ln\left(\frac{b}{a}\right)\mathrm{H\,m^{-1}} \tag{3.81}$$

Twin feeder

$$L' = \frac{\mu_0}{4\pi} + \frac{\mu_0}{\pi}\ln\left(\frac{d-a}{a}\right)\mathrm{H\,m^{-1}} \tag{3.82}$$

Microstrip

$$L' = \frac{\mu_0}{8\pi} + \mu_0\frac{h}{w}\mathrm{H\,m^{-1}} \tag{3.83}$$

or

$$L' = \frac{\mu_0}{8\pi} + \frac{\mu_0}{2\pi}\ln\left(\frac{h}{w}\right)\mathrm{H\,m^{-1}} \tag{3.84}$$

We also came across a very important relationship: the capacitance and inductance per unit length are related by

$$L'C' = \mu\varepsilon \tag{3.85}$$

Thus, we can find the inductance if we know the capacitance.

In common with capacitors, we found that inductors can also store energy. However, there is one major difference: a capacitor can store energy for a very long time, whereas the energy stored in a magnetic field disappears when the field collapses. We found that the stored energy can be given by

$$\text{energy} = \tfrac{1}{2}LI^2 \text{ J} \tag{3.86}$$

or

$$\text{energy} = \tfrac{1}{2}BH^2 \text{ J}\,\text{m}^{-3} \tag{3.87}$$

We were then introduced to the reactance of an inductor given by

$$X_L = 2\pi f L \tag{3.88}$$

with the supply voltage leading the coil current by 90°.

As with the last chapter, we could have examined parallel and series combinations of inductors. The reason why we didn't do this is that it is very easy to wind a coil on a former. Thus, non-standard value inductors are far simpler to produce than non-standard capacitors.

We concluded the chapter with a brief examination of plasma confinement in a fusion reactor. We saw that the magnetic field in use at the Joint European Torus fusion reactor is 3.4 Wb m^{-2}. This is needed to confine a 200 million°C plasma.

4 Electroconductive Fields

We now come to the last field system we are going to study – electroconductive fields. As electrical engineers, this is probably the most familiar field system in that every time we build an electrical circuit, we use electroconduction – charge flow around the circuit causes a current. However, what exactly is the nature of the current, how is it conducted and why does it flow? We can answer these questions by studying electroconductive fields.

4.1 CURRENT FLOW

When we considered electrostatics in Chapter 2, we saw that like charges repel each other. Electrical conductors put this effect to good use. To be a good conductor of electricity, the atoms in the material must have a large number of electrons in the conduction band. Most metals satisfy this requirement and are malleable enough to be drawn into wires.

Let us consider what happens when we connect a wire to a source of electrons, a battery for instance (see Figure 4.1). The battery produces electrons at the negative terminal by chemical means. So, when the negative terminal produces an electron, it repels electrons in the wire, so forcing an electron out of the wire into the positive battery terminal. We define the current through the wire as the rate of flow of charge, i.e.,

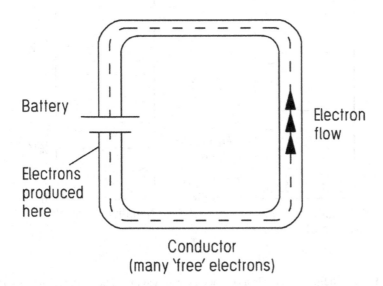

FIGURE 4.1   A conductor connected to a voltage source.

$$i = \frac{\mathrm{d}Q}{\mathrm{d}t} \qquad (4.1)$$

with units of coulomb per second, or ampere.

So, current is equal to the rate of flow of charge. This raises the question: what forces the charge around the circuit? This is where we reintroduce the ideas of potential and electric field strength.

4.2 POTENTIAL AND ELECTRIC FIELD STRENGTH

In Chapter 2, we came across potential as a measure of the work done against an electrostatic field. Now that we are considering an electroconductive field, what part does potential play?

Figure 4.2 shows a piece of conducting material connected across a battery. Now, the positive terminal of the battery will attract electrons, while the negative terminal will repel electrons. So, if we move a positive test charge from the negative terminal to the positive terminal, we have to do work against a field. The precise amount of work done is equal to the potential of the battery. (This is precisely the same definition of potential that we used when we considered electrostatics.)

As Figure 4.2b shows, we can draw lines of equal potential in the conductor. This diagram shows that the potential increases as we move towards the positive

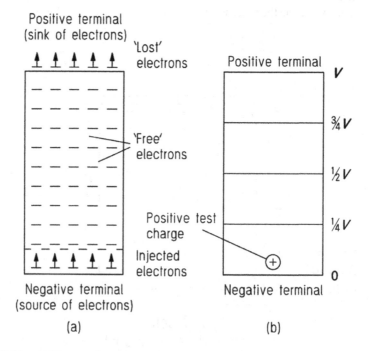

FIGURE 4.2   (a) Electron flow in a conductor connected to a voltage source and (b) equipotentials in a conductor.

terminal, confirming that we must do work against the field as we move our positive test charge along the conductor.

Let us now consider what happens to an electron injected into the conductor from the negative terminal of the battery. The positive terminal attracts the electron, while the negative terminal repels it. Thus, the electron experiences an electrostatic force that moves it up the conductor. The magnitude of this E field is given by

$$E = \frac{V}{l} \tag{4.2}$$

where l is the length of the conductor.

So, we have seen that a voltage source sets up an electric field in any conductor connected across it. We have also seen that the potential of the source is equal to the work done in moving a positive test charge around the external circuit. Charge flowing around the external circuit does so under the influence of the external field. The current represents this charge flow, and this is the subject of the next section.

4.3 CURRENT DENSITY AND CONDUCTIVITY

We have just seen that charges move around a circuit under the influence of an electric field. This charge flow can be regarded as a current by virtue of Equation (4.1):

$$i = \frac{dQ}{dt}$$

What precisely happens when the source injects an electron into the conductor? Well, this electron immediately enters an electric field that accelerates it towards the positive terminal. However, because the electron is travelling in a crystal lattice, it cannot travel very far before it meets an atom. Metal atoms have a large number of electrons available for conduction, and so, when electron meets a metal atom, the injected electron will dislodge a conduction band electron from the atom. The injected electron will then be caught by metal atom and the dislodged electron will travel through the lattice under influence of the electric field. Of course this new electron will soon meet a metal atom and the whole process repeats along the length of the conductor. Thus, the original electron takes a long time to reach the end of the wire, but the effect is felt after quite a short time. (In fact, the disturbance travels at the speed of light in the material surrounding the conductor.)

So, the original charge carriers from the battery cannot travel very far through the lattice before meeting a metal atom. Let us consider the motion of an electron across a volume of conductor (see Figure 4.3). These electrons will have an average velocity, known as the drift velocity, with symbol V m s^{-1}. Thus, in time δt, the electrons will have moved a distance $v\delta t$ metre.

If N is the number of electrons per unit volume, m^{-3}, and δs is the cross-sectional area of the volume, m^{-2}, the amount of charge passing through the area δs is

$$\delta Q = Nq \ v\delta t \ \delta s \ C \tag{4.3}$$

As current is the rate of flow of charge, the current through this area is

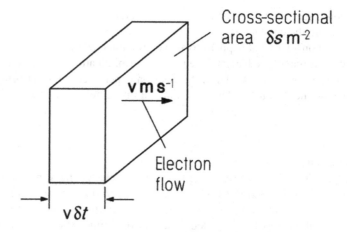

FIGURE 4.3   Elemental section through a current-carrying conductor.

$$i = \frac{\delta Q}{\delta t}$$

$$= Nqv \; \delta s \; \text{A} \qquad\qquad (4.4)$$

Thus, the current per unit area, or the current density, is

$$J = \frac{i}{\delta s}$$

$$= Nqv \; \text{A} \, \text{m}^{-2} \qquad\qquad (4.5)$$

We have reasoned that charges move through the conductor under the influence of an electric field. Thus, we can intuitively reason that the velocity of the electrons will depend on the value of E. Specifically, we can write

$$v = \mu_e E \, \text{m} \, \text{s}^{-1} \qquad\qquad (4.6)$$

where μ_e, is the mobility of the electrons in the lattice. By combining Equations (4.5) and (4.6), we get

$$J = Nq\mu_e \; E$$

or

$$J = \sigma \; E \qquad\qquad (4.7)$$

where σ is known as the conductivity of the material.

So, a material constant relates the current density to the electric field strength. This is identical in form to the corresponding relationships for electrostatics, $D = \varepsilon E$, and electromagnetism, $B = \mu H$.

According to Equation (4.7), the current density is directly dependent on the electric field strength. Thus, if we double the electric field strength, the current density will double. Of course, if the conductivity rises, the current density will also rise for a given E field. Thus, we can see that good conductors must have a large conductivity.

Example 4.1

A 12 V potential is set up across a conductor of length 50 cm. The conductor is made of copper, $\sigma = 58$ MS m^{-1}, and has a uniform cross-sectional area of 10 cm^2. Determine the electric field strength, the current density and the current that flows through the conductor.

Solution

The potential across the conductor is 12 V and the length is 50 cm. So, the electric field strength is

$$E = \frac{12}{50 \times 10^{-2}}$$

$$= 24 \text{ V m}^{-1}$$

Now,

$$J = \sigma E$$

and so,

$$J = 58 \times 10^6 \times 24$$

$$= 1.4 \times 10^9 \text{ A m}^{-2}$$

As the cross-sectional area of the sample is 10 cm^2, the current is

$$I = 1.4 \times 10^9 \times 10 \times 10^{-4}$$

$$= 1.4 \text{ MA}$$

It is a good job that this is simply an example, because such a large current would quickly vaporize the conductor!

4.4   RESISTORS

In the last section, we saw that the current density and electric field strength are related by (Equation (4.7))

$$J = \sigma E$$

Now, by substituting for J and E, we get

$$\frac{I}{A} = \sigma \frac{V}{l}$$

which, after some rearranging, gives

$$\frac{V}{I} = \frac{l}{\sigma A}$$

or

$$\frac{V}{I} = R \tag{4.8}$$

where R is the resistance of the conductor in ohms, given by

$$R = \frac{l}{\sigma A} \tag{4.9}$$

(An alternative expression for the resistance is

$$R = \frac{\rho l}{A} \tag{4.10}$$

where ρ is the resistivity of the material ($\rho = 1/\sigma$). In this book, however, we will use conductivity rather than resistivity.)

Equation (4.8) is known as Ohm's law, named after Georg Simon Ohm, 1789–1854, the German physicist who studied electroconductive fields. Stated simply, it says that the current flowing through a conductor is directly proportional to the voltage across it, and inversely proportional to the resistance. So, the lower the resistance, the higher the current.

In passing through any resistor, the current has to do work, and this generates heat. (A familiar example of this is the heat generated by an electric fire.) The amount of power dissipated by a resistor is given by

$$\text{power} = I^2 R \text{ W} \tag{4.11}$$

Example 4.2

A 12 V potential is set up across a conductor of length 50 cm. The conductor is made of copper, $\sigma = 58$ MS m^{-1}, and has a uniform cross-sectional area of 10 cm^2. Determine the resistance of the conductor, and calculate the heat generated.

Solution

The conductor has a regular cross-sectional area, and so the resistance is given by

$$R = \frac{l}{\sigma \times \text{area}}$$

$$= \frac{50 \times 10^{-2}}{58 \times 10^{6} \times 10 \times 10^{-4}}$$

$$= 8.6 \ \mu\Omega$$

The power dissipated by this resistance is

$$\text{Power} = I^2 R$$

$$= \left(\frac{12}{8.6 \times 10^{-6}}\right)^2 8.6 \times 10^{-6}$$

$$= 16.7 \ \text{MW}$$

So, this resistor is a very efficient producer of heat. Again, we must use care here because the heat generated is so high that it might vaporize the conductor.

Before we consider the resistance of the various transmission lines we have covered in previous chapters, let us examine the resistance of a capacitor.

4.4.1 CAPACITORS

At first sight it might seem that capacitors are out of place in a chapter on electro-conduction. However, we should remember that capacitors have a dielectric between the two plates, and this material can conduct a small amount of current – the leakage current.

If the capacitor has a regular cross section, we can find the resistance from

$$R = \frac{d}{\sigma A} \tag{4.12}$$

with the capacitance given by

$$C = \frac{\varepsilon A}{d} \tag{4.13}$$

So, a non-ideal capacitor can have both displacement current and conduction current between the two plates. The proportion of conduction current to displacement current is a measure of the loss of the capacitor.

Figure 4.4 shows the two currents drawn on an Argand diagram. As this figure shows, the conduction current is drawn on the real axis, and the displacement current is

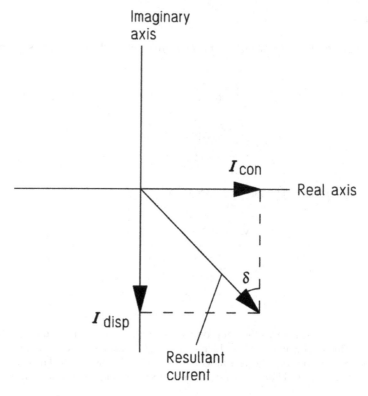

FIGURE 4.4 Relationship between conduction and displacement current in a non-ideal capacitor.

drawn on the negative imaginary axis. (It is drawn on the negative axis because it leads the conduction current by 90°.) Now, the conduction current is given by

$$I_{con} = \frac{V}{R}$$

$$= \frac{V \times \sigma \times \text{area}}{d} \tag{4.14}$$

And the displacement current is given by

$$I_{disp} = \frac{V}{X_C}$$

$$= \frac{V}{1/\omega C}$$

$$= V\omega \frac{\varepsilon \times \text{area}}{d} \tag{4.15}$$

As Figure 4.4 shows, the ratio of these two currents is also equal to the tangent of the angle δ. Thus,

$$\tan \delta = \frac{I_{con}}{I_{disp}}$$

$$= \frac{V \times \sigma \times area}{d} \times \frac{d}{V\omega\varepsilon \times area}$$

i.e.,

$$\tan \delta = \frac{\sigma}{\omega\varepsilon} \tag{4.16}$$

Equation (4.16) shows that $\tan \delta$ is a measure of the loss in the capacitor. Thus, $\tan \delta$ is known as the loss tangent of the capacitor. Although readers may wonder at the use of such a quantity, it is quite important as the following example shows.

Example 4.3

A parallel plate capacitor consists of two metal plates, of area 2 cm², separated by 3 μm of porcelain with $\varepsilon_r = 5.7$ and $\sigma = 2 \times 10^{-13}$ S m⁻¹. The capacitor is connected to a 12 V 50 Hz supply. Determine the conduction current, and compare it to the displacement current. In addition, calculate the loss tangent at frequencies of 50 Hz, 1 MHz and 100 MHz.

Solution

The capacitor is connected to a 12 V, 50 Hz supply. As the dielectric is lossy, the capacitor will also have some resistance given by

$$R = \frac{d}{\sigma \times area}$$

$$= \frac{3 \times 10^{-6}}{2 \times 10^{-13} \times 2 \times 10^{-4}}$$

$$= 7.5 \times 10^{10}\,\Omega$$

By applying Ohm's law, the leakage current is

$$I = \frac{12}{7.5 \times 10^{10}}$$

$$= 0.16 \text{ nA}$$

To find the displacement current, we need to find the capacitance. So,

$$C = \frac{\varepsilon_0 \varepsilon_r \times \text{area}}{d}$$

$$= \frac{8.854 \times 10^{-12} \times 5.7 \times 2 \times 10^{-4}}{3 \times 10^{-6}}$$

$$= 3.4 \text{ nF}$$

At 50 Hz, the reactance is

$$X_C = \frac{1}{2\pi f C}$$

$$= \frac{1}{2\pi \times 50 \times 3.4 \times 10^{-9}}$$

$$= 946 \text{ k}\Omega$$

and so the displacement current is

$$I = \frac{12}{946 \times 10^3}$$

$$= 12.7 \, \mu A$$

Thus, the displacement current is 79,000 times greater than the leakage current. This difference is quite fortunate, otherwise capacitors would be useless!
The loss tangent is given by (Equation (4.16))

$$\tan \delta = \frac{\sigma}{\omega \varepsilon}$$

and so

$$\tan \delta = \frac{2 \times 10^{-13}}{2\pi f \times 8.854 \times 10^{-12} \times 5.7}$$

$$= \frac{6.3 \times 10^{-4}}{f}$$

The following table compares the loss tangent at various frequencies.
Note that $\tan \delta$ is increasing with frequency. This is because the reactance of the capacitor is decreasing, so increasing the displacement current. The leakage current is relatively independent of frequency and so the lower the value of $\tan \delta$, the greater the insulating properties of the material. The converse is also true – the greater the loss tangent, the better the conduction properties of the material.

Frequency (Hz)	$\tan \delta$
50	1.3×10^{-5}
1×10^6	6.3×10^{-10}
100×10^6	6.3×10^{-12}

Before we leave the parallel plate capacitor, let us return to our expressions for capacitance and resistance, Equations (4.12) and (4.13). If we multiply these two together, we get

$$RC = \frac{d}{\sigma A} \frac{\varepsilon A}{d}$$

$$= \frac{\varepsilon}{\sigma} \tag{4.17}$$

Thus, it would appear that we have another fundamental relationship linking resistance and capacitance. However, this is only one example, and we should not start applying this 'rule' with enthusiasm just yet.

4.4.2 COAXIAL CABLE

With coaxial cable, we have two resistances to consider: the resistance of the inner conductor, and the resistance between the inner and outer conductors. As regards the first resistance, the cross section of the inner conductor is circular, and so the resistance is given by

$$R = \frac{l}{\sigma A} \Omega$$

or

$$R' = \frac{1}{\sigma A} \Omega \text{ m}^{-1} \tag{4.18}$$

Let us now turn to the resistance between the inner and outer conductors. If the inner conductor is at a potential of V volt above the outer conductor, there will be an electric field set up in the dielectric. In addition, leakage current will flow in a radial direction outwards from the inner conductor (see Figure 4.5). Now, the current density at a certain radius will be

$$J = \frac{I}{2\pi r \times \text{length}} \tag{4.19}$$

Thus, the E field at this radius is

$$E = \frac{J}{\sigma}$$

$$= \frac{I}{\sigma 2\pi r \times \text{length}} \tag{4.20}$$

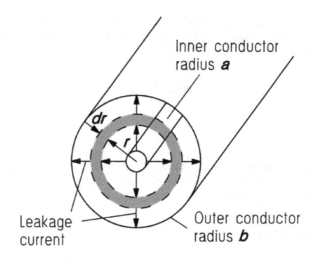

FIGURE 4.5 Leakage current in a length of coaxial cable.

Let us consider a thin, hollow tube of radius r and the thickness dr. The potential across this tube, dV, is

$$dV = \frac{-I}{\sigma 2\pi r \times \text{length}} dr$$

and so the total voltage between the inner and outer conductors is

$$\int_{0}^{v} dV = \int_{b}^{a} \frac{-I}{\sigma 2\pi r \times \text{length}} dr$$

Hence,

$$V = \frac{-I}{2\pi\sigma \times \text{length}} \int_{b}^{a} \frac{1}{r} dr$$

$$= \frac{-I}{2\pi\sigma \times \text{length}} \left. \ln r \right|_{a}^{b}$$

$$= \frac{-I}{2\pi\sigma \times \text{length}} \ln\left(\frac{b}{a}\right)$$

Thus, the leakage resistance of the cable is

$$R = \frac{-I}{2\pi\sigma \times \text{length}} \ln\left(\frac{b}{a}\right) \Omega \qquad (4.21)$$

or

$$R' = \frac{\ln\left(\dfrac{b}{a}\right)}{2\pi\sigma}\ \Omega\,m \tag{4.22}$$

As we have already seen, the capacitance of such a cable is given by, Equation (2.37),

$$C' = \frac{2\pi\varepsilon_0\varepsilon_r}{\ln\left(\dfrac{b}{a}\right)} \times F\,m^{-1}$$

So, the product of R' and C' is

$$R'C' = \frac{\ln\left(\dfrac{b}{a}\right)}{2\pi\sigma}\ \frac{2\pi\varepsilon_0\varepsilon_r}{\ln\left(\dfrac{b}{a}\right)}$$

$$= \frac{\varepsilon_0\varepsilon_r}{\sigma} \tag{4.23}$$

We obtained the same result when we multiplied the resistance and capacitance of a parallel plate capacitor together. So we appear to have a new rule that enables us to find the resistance of a field system if we know the capacitance.

Example 4.4

A 500 m length of coaxial cable has an inner conductor of radius 2 mm and an outer conductor of radius 1 cm. The conductivity is 3×10^{-4} S m^{-1}, while the conductivity of the inner core is 58 MS m^{-1}. Determine the series and shunt resistance of the cable.

Solution

The cable is 500 m long, and so we can use Equation (4.20) to find the shunt resistance. Thus,

$$R = \frac{\ln\left(\dfrac{b}{a}\right)}{2\pi\sigma \times \text{length}}$$

$$= \frac{\ln\left(1\times10^{-2}/2\times10^{-3}\right)}{2\pi\times3\times10^{-4}\times500}$$

$$= \frac{\ln 5}{0.94}$$

$$= 1.7\ \Omega$$

We could have obtained this result from a knowledge of the cable capacitance. This cable is the same as that used in Example 2.9 in Section 2.9. From this example, the cable capacitance is 86.4 nF, and so the shunt resistance is

$$R = \frac{\varepsilon_0 \varepsilon_r}{\sigma C}$$

$$= \frac{8.854 \times 10^{-2} \times 5}{3 \times 10^{-4} \times 86.4 \times 10^{-9}}$$

$$= 1.7 \ \Omega$$

As the inner conductor has a regular cross-sectional area, we can use Equation (4.9) to give

$$R = \frac{1}{\sigma A}$$

$$= \frac{500}{58 \times 10^6 \times \pi \times (2 \times 10^{-3})^2}$$

$$= 0.8 \ \Omega$$

So, the cable has a series resistance of 0.8 Ω and a shunt resistance of 1.7 Ω. We have already seen that this cable has inductance and capacitance as well, and we must take these into account when determining whether the cable is useful for a particular purpose.

4.4.3 TWIN FEEDER

Let us now consider twin feeder. This type of cable is usually air spaced and so the shunt resistance is zero. (If air conducted electricity, our electricity bills would be even higher than now!) So, air-spaced twin feeder only has the series resistance of the conductors. Thus,

$$R = \frac{l}{\sigma A} \ \Omega \qquad\qquad (4.24)$$

or

$$R' = \frac{l}{\sigma A} \ \Omega \ \mathrm{m}^{-1} \qquad\qquad (4.25)$$

Of course, if the twin feeder were not air spaced but surrounded by a lossy dielectric, the shunt resistance would be given by

$$R = \frac{\ln\left(\dfrac{d-a}{a}\right)}{\pi \sigma \times \mathrm{length}} \ \Omega \qquad\qquad (4.26)$$

where we have made use of our new relationship Equation (4.23).

4.4.4 MICROSTRIP LINES

We seldom need to consider the shunt resistance of a length of microstrip. This is because printed circuit boards are usually made of PTFE or Teflon, $\sigma \approx 10^{-16}$ S m^{-1}, and this makes them very poor conductors. Instead, the voltage drop along a length of track is very important.

The metallization on printed circuit boards usually has a uniform cross-sectional area, and so the resistance of a length of track is given by

$$R = \frac{l}{\sigma A} \Omega \qquad (4.27)$$

where A is the cross-sectional area of the track. As copper is usually used on pcbs, the resistance per cm is

$$R' = \frac{1}{58 \times 10^8 \times w \times t}$$

$$= \frac{1.7 \times 10^{-10}}{w \times t} \Omega\,\text{cm}^{-1} \qquad (4.28)$$

where w is the width of the track, and t is the thickness of the metalization.

Example 4.5

A printed circuit board has a 10 cm long, 5 V supply rail etched on to it. This track feeds two logic devices, one of which is half-way along the line while other is at the end of the track. If the first device takes a current of 300 mA while the second only takes 10 mA, determine the supply voltage at both devices. If the second device suddenly takes 300 mA, determine the new supply voltages. Assume that the copper metallization is 20 μm thick and that the track width is 2 mm.

Solution

We have a 10 cm length of track feeding 5 V to two logic circuits. The first is half-way along the track and takes a current of 300 mA. As the track has a uniform cross-sectional area, and it is made of copper, the resistance per cm is given by (Equation (4.28))

$$R' = \frac{1.7 \times 10^{-10}}{w \times t} \Omega\ \text{cm}^{-1}$$

$$= \frac{1.7 \times 10^{-10}}{2 \times 10^{-3} \times 20 \times 10^{-6}} \Omega\ \text{cm}^{-1}$$

$$= 4.25\,\text{m}\Omega\ \text{cm}^{-1}$$

Now, the first device is 5 cm from the source. So, the resistance of the first length is

$$R = 4.25 \times 10^{-3} \times 5$$

$$= 21.25 \text{ m}\Omega$$

The current flowing through this track is the sum of the individual currents, i.e.,

$$I = 300 + 10 \text{ mA}$$

$$= 310 \text{ mA}$$

So, the voltage drop is

$$V = IR$$

$$= 310 \times 10^{-3} \times 21.25 \times 10^{-3}$$

$$= 6.6 \text{ mV over the first section}$$

Thus, the first device has a supply voltage of $V = 5$–$6.6 \times 10^{-3} = 4.9934 \text{ V}$

$$V_1 = 5 - 6.6 \times 10^{-3}$$

$$= 4.9934 \text{ V}$$

The current through the second section of the line is 10 mA. Thus, the voltage drop across this section is

$$V = IR$$

$$= 10 \times 10^{-3} \times 21.25 \times 10^{-3}$$

$$= 0.213 \text{ mV}$$

and so the second device has a supply voltage of

$$V_2 = 4.9934 - 0.213 \times 10^{-3}$$

$$= 4.9932 \text{ V}$$

By following a similar procedure, the supply voltages when the second device takes 300 mA are

$$V_1 = 4.98725 \text{ V}$$

and

$$V_2 = 4.98088 \text{ V}$$

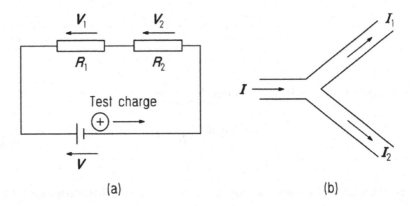

FIGURE 4.6 (a) Series connection of two resistors and (b) current split between two conductors.

Although the difference in supply rails is small in this example, it does show the dangers of making the supply rail too thin. We can also see the effect of switching on a logic gate: the power rail will drop for all devices supply the line. If the logic devices are turned on and off, as in a sequential logic design, this can lead to sudden voltage variations throughout the board. The solution is to use decoupling capacitors and wide tracks. (Decoupling capacitors are placed between the supply and ground very close to the logic devices. As the capacitors are charged to the supply rail, they can supply any sudden demand for current. Thus, the voltage along the power rail will not suddenly drop.)

4.4.5 KIRCHHOFF'S VOLTAGE AND CURRENT LAWS

While we are considering electroconductive fields, let us take some time to examine Kirchhoff's laws which are more often treated as 'circuit' laws. Figure 4.6a shows a circuit in which two resistors are connected in series. This combination is connected to a source of potential V. Now, what happens as we move a positive test charge from the negative terminal to the positive terminal via the external circuit?

Let us move the test charge from the negative terminal through R_2 to junction between the two resistors. As there is an E field set up in resistor R_2 we have to do work against the field. The magnitude of the E field can be written as

$$E_2 = \frac{V_2}{l_2} \tag{4.29}$$

where l_2 is the length of resistor R_2. Now, the force on our unit test charge is

$$F_2 = \frac{V_2}{l_2} \times 1$$

$$= \frac{V_2}{l_2} \text{N}$$

and so the work done in moving through the resistor is

$$\text{Work done} = \frac{V_2}{l_2} \times l_2$$

(4.30)

$$= V_2$$

Similarly, the work done in moving through resistor R_1 is

$$\text{Work done} = \frac{V_2}{l_2} \times l_1$$

(4.31)

$$= V_1$$

Hence, the total work done in moving the test charge around the external circuit is

$$\text{Total work done} = V_1 + V_2$$

(4.32)

This total work done must be equal to the potential of the source. (Any shortfall in the potential would be dropped across another resistance, and so we would have to do additional work against another field.) Thus,

$$V = V_1 + V_2$$

(4.33)

i.e., the sum of the potential drops around an external circuit is equal to the supply potential. This is Kirchhoff's voltage law. So, the potential difference around a circuit must add up to the supply potential. However, what happens to the current at a junction in an external circuit?

Figure 4.6b shows a current-carrying conductor splitting into two paths. Let us consider a current flowing through the left-hand conductor. This current will split between the two right-hand conductors, with the resistance of the two branches determining the exact proportions of the split. Now, the only current entering the junction, I, is from the left-hand conductor, and the only currents leaving the junction, I_1 and I_2, do so through the right-hand conductors. As there are no other sources of charge and no sinks of charge, we can reason that the current entering the junction is equal to the current leaving. Thus,

$$I = I_1 + I_2$$

(4.34)

In general, we can state that the current entering a junction must equal the current leaving. This is Kirchhoff's current law which we can regard as an example of 'good housekeeping' – we have to account for all sources and sinks of current. A similar situation exists in all the field problems we have met – the flux entering a surface must equal the flux leaving the surface if the surface does not enclose a source of flux.

4.4.6 COMBINATIONS OF RESISTORS

Like capacitors, resistors come in standard values. Sometimes this can be very inconvenient, and so we have to make our own values. One way to do this is to trim an existing resistor by gently filing away at the body! This requires a very steady hand and a great deal of patience. An alternative is to combine resistors in series or parallel.

Figure 4.7a shows two resistors, R_1, and R_2, in parallel. We require to find the equivalent resistance of this arrangement. Let us connect a d.c. source, V_s, to the resistors. Now, both resistors have the same voltage across them, but they will pass different currents. So,

$$I_1 = \frac{V_s}{R_1} \tag{4.35a}$$

and

$$I_2 = \frac{V_s}{R_2} \tag{4.35b}$$

If we replace the two resistors by a single equivalent one of value R_t, the current taken by the new resistor, I_t, must be the same as that taken by parallel combination. Thus,

$$I_t = \frac{V_s}{R_t} \tag{4.36}$$

To be equivalent, I_t must be the sum of the individual currents. So,

$$I_t = I_1 + I_2$$

or

$$\frac{V_s}{R_t} = \frac{V_s}{R_1} + \frac{V_s}{R_2}$$

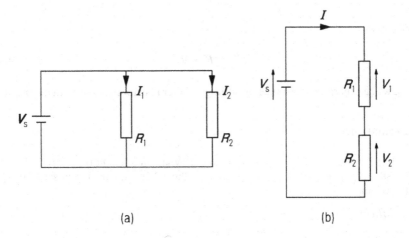

(a)   (b)

FIGURE 4.7   (a) Parallel combination of two resistors and (b) series combination of two resistors.

Thus,

$$\frac{1}{R_t} = \frac{1}{R_1} + \frac{1}{R_2} \tag{4.37}$$

So, we can decrease resistance by adding another resistor in parallel with the original.

Let us now consider a series combination of resistors as shown in Figure 4.7b. As before, we will connect this combination to a d.c. source and replace the resistors by a single equivalent one. Now, each of the individual resistors will have a potential across them, but pass the same current. Thus,

$$V_1 = IR_1 \tag{4.38a}$$

and

$$V_2 = IR_2 \tag{4.38b}$$

These individual potentials will add to give the supply voltage. So,

$$V_s = V_1 + V_2 \tag{4.39}$$

However, our equivalent resistor will have a potential across it given by

$$V_s = IR_t \tag{4.40}$$

So, by combining Equations (4.40), (4.39) and (4.38a and b), we get

$$IR_t = IR_1 + IR_2$$

resulting in

$$R_t = R_1 + R_2 \tag{4.41}$$

Thus, we can increase resistance by adding another resistor in series with the original.

Example 4.6

A 100 kΩ resistor is connected in a circuit. What is the effect of placing a 1 kΩ resistor in parallel with it? What happens if the 1 kΩ is connected in series with the original?

Solution

We have a 100 kΩ resistor and a 1 kΩ connected in parallel. So, the total

$$\frac{1}{R_t} = \frac{1}{R_1} + \frac{1}{R_2}$$

$$= \frac{1}{100} + \frac{1}{1}$$

$$= 1.01$$

i.e.,

$$R_t = 0.99 \text{ k}\Omega$$

So the 1 kΩ resistor dominates the 100 kΩ when they are in parallel. If we now connect the 1 kΩ resistor in series, we get a new resistance of

$$R_1 = 100 + 1$$

$$= 101 \text{ k}\Omega$$

Thus, the resistance has barely altered at all.

4.5   SOME APPLICATIONS

In general, electroconduction is the most familiar of all the field systems we have considered. Some typical applications include resistors, conducting wire, fuse wire and heating elements in electric fires, cookers and immersion heaters. As these examples are so familiar, we will not consider them. Instead, we will examine the formation of a resistor in semiconductor material.

When designing analogue circuits, resistors are often used. If the circuits are fabricated on printed circuit board, the designer can choose between wire-ended or surface-mount resistors. When producing an analogue integrated circuit, designers have no choice but to fabricate the resistors out of semi-conductor material.

Figure 4.8 shows the basic form of an integrated circuit (i.c.) resistor. It is formed by depositing electrodes on to the silicon and doping the material to the required conductivity. (Doping involves diffusing impurities into the silicon to alter the material conductivity.)

When working with i.c. resistors, the surface resistance of the material is used. If we consider a square of material, with electrodes on opposite sides, the resistance will be given by

$$R = \frac{l}{\sigma \times \text{area}}$$

where l is the length of one side of the square, σ is the material conductivity and area is the cross-sectional area of the resistor.

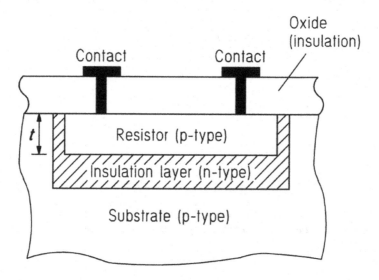

FIGURE 4.8   A basic integrated circuit resistor.

Thus,

$$R = \frac{l}{\sigma \times l \times t}$$

$$= \frac{1}{\sigma \times t} \tag{4.42}$$

where t is the thickness of the resistor. So, the resistance of a square of material is independent of the surface area. This is a very important result because it means that i.c. designers can reduce the size of their resistors without altering the resistance.

The precise resistance per square depends on the impurity diffusion and thickness, but usually ranges from 2 to 200 Ω sq^{-1}. So, if we have resistor with dimensions 1 by 10, we have ten squares giving a resistance of 20 Ω to 2 kΩ.

4.6   SUMMARY

This chapter has been concerned with probably our most familiar field systems – electroconductive fields. We began by examining what we mean by current flow. We then went on to discuss the use of potential and electric field strength. This led us on to a discussion of current density and introduced us to the fundamental relationship

$$J = \sigma E \tag{4.43}$$

We were then introduced to Ohm's law,

$$V = \frac{I}{R} \tag{4.44}$$

and the idea of resistance,

$$R = \frac{l}{\sigma A} \qquad (4.45)$$

The resistance of a parallel plate capacitor, and various transmission lines, was then examined. The relevant resistances are summarized here:

Capacitor

$$R = \frac{d}{\sigma A} \, \Omega \, (\text{shunt}) \qquad (4.46)$$

Coaxial cable

$$R = \frac{d}{\sigma A} \, \Omega \, (\text{series}) \qquad (4.47)$$

and

$$R = \frac{\ln\left(\dfrac{b}{a}\right)}{2\pi\sigma \times \text{length}} \, \Omega \, (\text{shunt}) \qquad (4.48)$$

Twin feeder

$$R = \frac{l}{\sigma A} \, \Omega \, (\text{series}) \qquad (4.49)$$

Microstrip

$$R = \frac{l}{\sigma A} \, \Omega \, (\text{series}) \qquad (4.50)$$

We also came across a very important relationship: the capacitance and resistance are related by

$$RC = \frac{\varepsilon_0 \varepsilon_t}{\sigma} \qquad (4.51)$$

Thus, we can find the resistance if we know the capacitance.

One of the surprising aspects of this work is that a capacitor can also have resistance. This is due to imperfect dielectrics. (In reality, all dielectrics will allow some current to flow.) The loss tangent characterizes this loss as the ratio of conduction to displacement current, i.e.,

$$\tan\delta = \frac{I_{\text{con}}}{I_{\text{disp}}}$$

or

$$\tan\delta = \frac{\sigma}{\omega\varepsilon}$$

So, the lower the value of loss tangent, the better the dielectric. An alternative viewpoint is that the higher the proportion of displacement current, the better the dielectric.

Next, we examined parallel and series combinations of resistors. We saw that resistance can be decreased by adding another resistor in parallel with the original, and increased by adding series resistance.

We concluded this chapter with an examination of resistance in integrated circuits. We found that the resistance depends on the number of 'squares' between the resistor contacts. Thus, quite large resistors can be physically very small.

5 Comparison of Field Equations

We have now completed our study of electrostatic, electromagnetic and electroconductive fields. In this chapter, we will compare and contrast the three field systems we have met. This should prove instructive and reveal some interesting facts. The next two chapters discuss the physical effects of dielectrics in capacitors and iron cores on transformers.

We will begin our comparison by examining the creation of a force field in the three systems we have considered.

5.1 FORCE FIELDS

In all the field systems we have considered, we started from the basic premise that like charges or poles repel and unlike charges or poles attract. This led to the observation that the force decreases as the square of the distance between the charges or poles, i.e.,

Electrostatics

$$F = \frac{q_1 q_2}{4\pi\varepsilon r^2}$$

Magnetostatics

$$F = \frac{p_1 p_2}{4\pi\mu r^2}$$

Electroconduction

$$F = \frac{q_1 q_2}{4\pi\varepsilon r^2}$$

As we can see, there is a great deal of similarity in the general forms of these equations – they all describe a force field of some sort. We should note that these equations assume point sources and, as we have seen with magnetism, this can be open to question. In spite of this fact, this does help us to visualize the field systems.

Once we accept that we are studying a force field, it is a simple step forward to define field strength as the force on a unit charge or pole. Thus,

Electrostatics

$$E = \frac{q_1}{4\pi\varepsilon r^2}$$

Magnetostatics

$$H = \frac{p_1}{4\pi\mu r^2}$$

Electroconduction

$$E = \frac{q_1}{4\pi\varepsilon r^2}$$

Let us take a moment to study these equations in detail. We should be fairly happy with the force field in an electrostatic system – we have all rubbed a balloon and felt the hairs on the back of our hands stand up. Anyone who has played with permanent magnets has seen the effect of the force field in a magnetic system. However, the idea of a force field in electroconductive systems is rather difficult to accept. This is because we are generally taught, at an early stage, a model of current flow that has the potential of a source as the force that drives current around a circuit. This idea is often reinforced by the term electromotive force for the potential of a source. (A water analogy is also often used. Although this analogy can be useful, its pull is very seductive and we must be very careful.) As this model is very simple, it is generally very hard to accept that it is wrong. Potential does not force current around a circuit: it is the electric field set up in a conductor that causes current to flow.

So, each field system has a force field that repels or attracts charges or poles. This raises the question of what radiates from the charge or pole generating the force field. This is where we introduce the ideas of flux and flux density.

5.2   FLUX, FLUX DENSITY AND FIELD STRENGTH

In Chapter 2, we came across flux as the flow of material from one place to another. So, if a charge radiates flux, it should run out at some time! (After all, if material transfers from one place to another, the source will eventually run out.) This is where we must exercise extreme caution – flux is only a construct to help us visualize the field system. In electrostatic and magnetostatic field systems, no flux physically flows. In spite of this statement, in an electroconductive field system, the flow of flux does signify the transfer of charge and that is a physical phenomenon. (Well, there's always an exception to the rule!) So, a source radiates flux, and a sink attracts it. A little thought shows that the flux depends on the force field – a large force field implies a large flux. Let us now examine this in greater detail.

In all the field systems we considered, we met a flux density of some sort: electric flux density in electrostatics, magnetic flux density in electromagnetism and current density in electroconduction. In each of these systems, material constants related the flux densities to the respective field strengths, i.e.,

Electrostatics

$$D = \varepsilon E$$

where

$$D = \frac{\psi}{\text{area}} \text{ and } E = \frac{V}{l}$$

Electromagnetism

$$B = \mu H$$

where

$$B = \frac{\phi}{\text{area}} \text{ and } H = \frac{NI}{l} = \frac{V_m}{l}$$

Electroconduction

$$J = \sigma E$$

where

$$J = \frac{I}{\text{area}} \text{ and } E = \frac{V}{l}$$

This comparison implies a fundamental relationship:

$$\text{flux density} = \text{material constant} \times \text{field strength}$$

This is a quite remarkable result. Here we are with three different field systems, all of which have the same simple fundamental relationship. We should also note that these equations are spatial in form – they use the flux density, not the actual flux.

We can also observe that the field strength (or potential gradient) is given by

$$\text{field strength} = \frac{\text{potential}}{\text{path length}}$$

where potential is the work done in moving a unit test charge or pole around the path length. So, the spatial equations are all similar in form, but what about the 'point forms' of these equations?

5.3 POTENTIAL AND RESISTANCE TO FLUX

In the last section, we compared the three fundamental laws governing electrostatic, electromagnetic and electroconductive fields. We can produce the point form of these laws by substituting for the flux densities and field strengths to give, after some rearranging,

Electrostatics

$$V = \frac{\psi}{C}$$

where

$$\frac{1}{C} = \frac{1}{\varepsilon \times \text{area}}$$

Electromagnetism

$$V_m = \phi S$$

where

$$S = \frac{l}{\mu \times \text{area}}$$

Electroconduction

$$V = IR$$

where

$$R = \frac{l}{\sigma \times \text{area}}$$

Again there is a large degree of similarity among the three field systems. In particular, we can say

$$\text{potential} = \text{flux} \times \text{resistance to flux}$$

or

$$\text{work done} = \text{flux} \times \text{resistance to flux}$$

where the potential is the work done in moving a unit charge or pole around a path in a force field.

Also, the resistance to the flow of flux in all the systems is governed by the length of circuit:

$$\text{resistance to flux} = \frac{\text{length of circuit}}{\text{material constant} \times \text{area}}$$

5.4   ENERGY STORAGE

Of the three field systems we have considered, only the electrostatic and electromagnetic fields can store energy – the electroconductive system dissipates energy as current flows. From a circuits point of view, the stored energy is given by

Electrostatics

$$\text{energy} = \frac{1}{2}CV^2 \text{ J}$$

Electromagnetism

$$\text{energy} = \frac{1}{2}LI^2 \text{ J}$$

Now, capacitors rely on the voltage across them to generate an electric field, whereas inductors use the current through them to produce a magnetic field. With this in mind, we can see a great deal of similarity between the two circuit elements. This similarity is even greater if we compare the energy stored in the fields

Electrostatics

$$\text{energy} = \frac{1}{2}DE \text{ J m}^{-3}$$

Electromagnetism

$$\text{energy} = \frac{1}{2}BH \text{ J m}^{-3}$$

So, in general we can say:

$$\text{energy density} = \frac{1}{2} \times \text{flux density} \times \text{field strength}$$

Let us just take a moment to compare the efficiency of both methods of storing energy. Let us consider a capacitor made of two circular plates, of area $5\,cm^2$, separated by 3 µm of glass with ε, of 10. If the potential between the plates is $100\,V$, the stored energy is

$$\text{energy} = \frac{1}{2}\frac{14.8 \times 10^{-19}}{5 \times 10^{-4}}\frac{\times 100^2}{\times 3 \times 10^{-6}}$$

$$= 49 \text{ kJ m}^{-3}$$

In order to attain the same energy density, the H field in an inductor with a mumetal core, $\mu_r = 1 \times 10^5$, would need to be

$$H^2 = \frac{2 \times 49 \times 10^4}{4\pi \times 10^{-7} \times 1 \times 10^5}$$

and so,

$$H = 883 \text{ At m}^{-1}$$

If we take a coil length of 10 cm and pass 100 mA through the coil, this implies 883 turns of wire. If we assume that the capacitor and inductor are air cored, the inductor would need 88.3×10^3 turns for the same length! Obviously capacitors are better at storing energy.

5.5 FORCE

Both the electrostatic and magnetic field systems can be used to apply force to charged or magnetic surfaces. We were able to calculate the force by using the energy stored per unit volume and moving one of the surfaces a very small amount. This gave us the following results:

Electrostatics

$$\text{force} = \frac{1}{2}\psi E \, \text{N}$$

Electromagnetism

$$\text{force} = \frac{1}{2}\phi H \, \text{N}$$

Thus, we can say

$$\text{force} = \frac{1}{2} \times \text{flux} \times \text{field strength}$$

As with the previous section, let us compare the relative efficiencies of electromagnetic and electrostatic lifting equipment. Let us initially consider a horseshoe-shaped piece of iron of circular cross-sectional area 10 cm² and overall length 60 cm. To convert this to an electromagnet, two coils of 750 turns each are wound on each limb and connected in series. Let us try and find the current required to lift 200 kg masses, if an air gap of 0.1 mm is present at each pole.

Now, first we must find the force required to lift these weights. This lifting force must equal that due to gravity. So,

$$F = ma$$

$$= 200 \times 9.81$$

$$= 1.962 \times 10^3 \, \text{N}$$

This force must be generated by the electromagnet, and so,

$$1.962 \times 10^3 = \frac{1}{2}\phi H \times 2$$

$$= \frac{1}{2} B \times \text{area} \times \frac{B}{\mu_0} \times 2$$

(The factor of two appears in this equation because we effectively have two pole faces.) Thus,

$$B^2 = \frac{1.962 \times 10^3 \times 4 \times \pi \times 10^{-7}}{10 \times 10^{-4}}$$

$$= 2.466$$

or

$$B = 1.57 \text{ Wb m}^{-2}$$

This is the flux density in the air gap as well as that in the former of the electromagnet. If we take a μ_r of 410, we get a magnetic field strength of 3×10^3 At m^{-1} in the former. If we ignore the reluctance of the air gap and the metal being lifted, we get

$$H = \frac{NI}{l}$$

and so,

$$NI = 3 \times 10^3 \times 60 \times 10^{-2}$$

$$= 1.8 \times 10^3 \text{ At}$$

Thus, the current through the two coils of 750 turns each is

$$I = \frac{1.8 \times 10^3}{2 \times 750}$$

$$= 1.2 \text{ A}$$

This is a quite respectable current and one that a voltage source can supply without too much trouble.

Let us now turn our attention to the use of electrostatics. To make a fair comparison, we will assume that we have two metal plates of diameter equal to the spacing between the poles of the electromagnet. The use of a steel rule, and a little imagination, means we can take a plate diameter of 25 cm. If we assume the plates are separated by 0.1 mm as before, and we wish to lift 200 kg again, we get

$$1.962 \times 10^3 = \frac{1}{2} \psi E$$

$$= \frac{1}{2} \varepsilon_0 E \times \text{area} \times E$$

and so,

$$E^2 = 9 \times 10^{15}$$

or

$$E = 9.5 \times 10^7 \, \text{V m}^{-1}$$

This results in a potential between the plates of

$$V = 9.5 \times 10^7 \times 0.1 \times 10^{-3}$$

$$= 9.5 \, \text{kV}$$

This is a rather large voltage to say the least! In reality, air breaks down at a field strength of $3 \times 10^6 \, \text{V m}^{-1}$ and so we can't use electrostatics here.

So, to summarize, electrostatic fields are more efficient at storing energy than electromagnetic fields. However, electromagnetic fields are more efficient at lifting materials.

5.6 SUMMARY

We have seen a great deal of commonality among our three different field systems. The fundamental relationships we found are summarized here:

$$\text{flux density} = \text{material constant} \times \text{field strength}$$

$$\text{field strength} = \frac{\text{potential}}{\text{path length}}$$

$$\text{potential} = \text{flux} \times \text{resistance to flux}$$

$$\text{resistance to flux} = \frac{\text{length of circuit}}{\text{material constant} \times \text{area}}$$

$$\text{energy density} = \frac{1}{2} \times \text{flux density} \times \text{field strength}$$

$$\text{force} = \frac{1}{2} \times \text{flux} \times \text{field strength}$$

We have now finished our comparison of our three field systems. In the next chapter, we will examine what happens to dielectrics in an electric field.

6 Dielectrics

In this chapter, we will re-examine capacitors and, in particular, the effect of an electric field on dielectrics. As we saw in Chapter 2, dielectrics have a permittivity greater than air. Thus, when we use them as the insulating material in a capacitor, they increase the capacitance. If the potential across the plates is constant, the stored charge will increase and so it is generally desirable to design a capacitor with a dielectric.

The next section introduces the dipole moment that occurs when an electric field distorts an atom, or molecule.

6.1 ELECTRIC DIPOLES AND DIPOLE MOMENTS

If we have an atom that is well away from an external electric field, the orbit of the electrons will describe a sphere with the nucleus at the centre (see Figure 6.1a). If we place this atom in an electric field, the field will distort the atom as Figure 6.1b shows. In effect, the electron cloud moves away from the field, whereas the positively charged nucleus moves in the direction of the field. The atom is said to be polarized by the electric field. Although the polarized atom is still electrically neutral, on the microscopic scale the atom has an electric field. This becomes clearer if we replace the electron cloud by a point source at a distance d from the positive nucleus. This arrangement, shown in Figure 6.1c, is known as an electric dipole.

So, when we subject an electrically neutral atom to an external electric field, we can represent the atom by an electric dipole. We tend to characterize electric dipoles by the dipole moment, p, given by

$$p = q \times d \ C \ m \tag{6.1}$$

where q is the charge on the nucleus or electron cloud and d is the effective separation of the two charges. The dipole moment is a vector quantity with direction towards the positive charge. The reason for this choice of direction will become clear in the next section. In most materials, the separation between the charges is directly dependent on the magnitude of the external E field. However, if we continually increase the E field, there comes a point where we cannot polarize the atom anymore. The material is then said to be saturated. If we increase the field beyond this point, the force on the dipole becomes so great that it breaks apart and becomes ionized. If the dielectric is the insulating material in a capacitor, this causes a small, but very dramatic, explosion! So, it is important to know the field at which the dielectric breaks down. This is the subject of Section 6.4.

Some materials exhibit dipole moments on a molecular level, or macroscopic scale. Figure 6.2 shows a charge embedded in some insulating material. As this charge is negative, it tends to polarize the surrounding molecules producing molecular dipole moments.

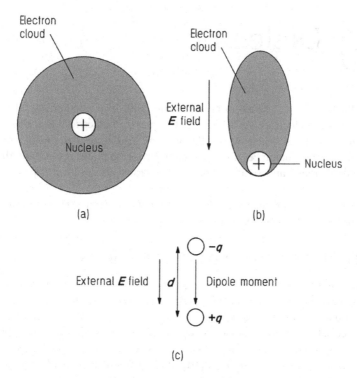

FIGURE 6.1   (a) Structure of an atom with no external E field, (b) polarization of atom due to external E field and (c) equivalent electric dipole.

The net effect is that these molecular dipoles are able to exert a force on each other, and on any external charges.

Now, what happens if we introduce a test charge into this region? Logically each dipole moment will exert a force on the test charge. However, the net force will be less than if the material was not there. This is because the force acting on the test charge is due to the vector sum of the dipole moments. As the distribution of these dipoles is random throughout the material, the net result is a smaller force than if the material was not present.

An alternative explanation is to consider the direction of the electric field produced by each dipole. As Figure 6.2 shows, these fields act in the opposite direction to the point charge field, so reducing the field from the point charge. We can check the validity of this model by noting

$$E = \frac{D}{\varepsilon} \tag{6.2}$$

So, if the charge is in air, the force field is

$$E = \frac{D}{\varepsilon_0} \tag{6.3}$$

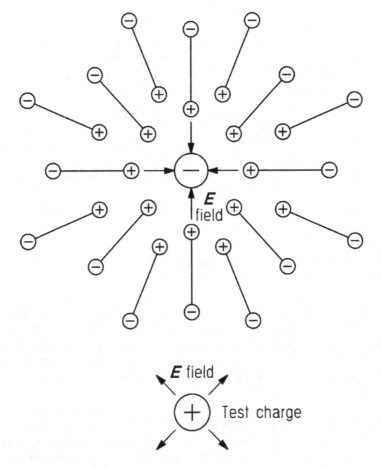

FIGURE 6.2   Effect of charge on molecular electric dipoles.

However, if the charge is in material with a relative permittivity of ε_r, the force field is

$$E = \frac{D}{\varepsilon_0 \varepsilon_r}$$

(6.4)

which is lower than for air. This implies that it is the polarization of the material that causes a permittivity greater than air. This is the subject of the next section.

6.2   POLARIZATION AND RELATIVE PERMITTIVITY

We have just seen that the material in which a charge is placed reduces the electric field strength produced by a point charge. What we have not yet examined is the effect of a dielectric on a capacitor.

Figure 6.3a shows a capacitor with a dielectric between its plates. When the upper plate has a positive charge on it, flux radiates down across the air gap and

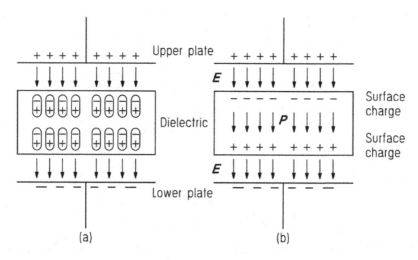

FIGURE 6.3 (a) Dielectric-filled capacitor and (b) macroscopic view of dielectric-filled capacitor.

into the dielectric. The E field in the capacitor will polarize the dielectric, so producing a negative charge on the upper surface of the dielectric. A similar situation occurs on the lower plate, and so we can take the macroscopic view as Figure 6.3b shows. So, the external field induces polarization in the dielectric with the dipole moments pointing towards the lower plate. This implies a polarization vector which we can define as follows.

Let us consider a small volume of dielectric, δV, that is large enough to enclose a large number of atoms. If the number of atoms per unit volume is N, the number of atoms in this volume is $N\delta V$. Now, let us assume that all of these atoms are polarized in the same direction – acting downwards. Each atom has a dipole moment of

$$p = q \times d \ C \ m \tag{6.5}$$

and so the total polarization of the volume δV is

$$N\delta V \times qd \ C \ m \tag{6.6}$$

Thus, the polarization per unit volume, P, is

$$P = \frac{\text{polarization of } \delta V}{\delta V}$$

$$= \frac{N\delta V \times qd}{\delta V}$$

which gives

$$P = Nqd \ C \, m^{-2} \tag{6.7}$$

So, we have a new vector, the polarization, with units of coulomb per m² acting in the same direction as the E field. This means we can modify our equation linking D and E to give

$$D = \varepsilon_0 E + P \qquad (6.8)$$

Materials in which the dipole moment is directly proportional to the E field are known as linear materials, and most dielectrics we come across are linear in form. Thus, we can write

$$P = \chi \varepsilon_0 E \qquad (6.9)$$

where χ is a constant of proportionality called the electric susceptibility of the material. We can now write Equation (6.8) as

$$D = \varepsilon_0 E + \chi \varepsilon_0 E$$
$$= (1 + \chi) \varepsilon_0 E$$

Thus, the relative permittivity of the material is

$$\varepsilon_r = 1 + \chi \qquad (6.10)$$

i.e., ε_r, depends on the electric susceptibility of the material, which is itself a measure of the polarizability of the dielectric.

Example 6.1

Two metal plates of area 10 cm are placed 5 mm apart. A dielectric, with $\varepsilon_r = 2.5$ and thickness 3 mm, is placed mid-way between the plates. A potential of 200 V is applied across the plates. Determine:

1. the capacitance of the arrangement
2. the surface charge densities
3. the flux density at all points
4. the electric field strength at all points
5. the bound charge densities

Solution

Figure 6.4a shows the situation we seek to analyze. As this figure shows, there is an air gap both above and below the dielectric. Now, by considering the charge distribution throughout the capacitor, we can see that we can replace the composite capacitor by three capacitors in series (Figure 6.4b).

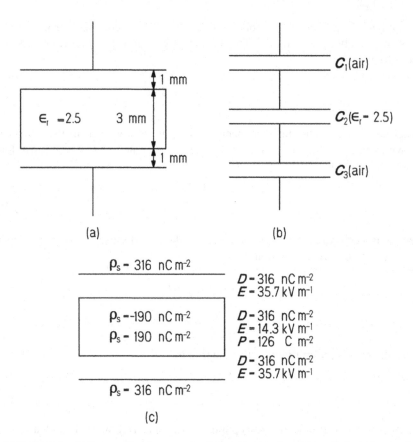

FIGURE 6.4   (a) Schematic of composite dielectric capacitor, (b) equivalent circuit of composite dielectric capacitor and (c) distribution of flux density and surface charges.

We can find the total capacitance from

$$\frac{1}{C_t} = \frac{1}{C_1} + \frac{1}{C_2} + \frac{1}{C_3}$$

$$= \frac{1 \times 10^{-3}}{\varepsilon_0 \times 10 \times 10^{-4}} + \frac{2 \times 10^{-3}}{\varepsilon_0 \times 2.5 \times 10 \times 10^{-4}} + \frac{1 \times 10^{-3}}{\varepsilon_0 \times 10 \times 10^{-4}}$$

and so,

$$C_t = 3.16 \, \text{pF}$$

The voltage across the capacitor is 100 V and so the stored charge is

$$Q_t = 3.16 \times 10^{-12} \times 100$$

$$= 316 \, \text{pC}$$

Thus, the top plate will have a charge of 316 pC on it giving a surface charge density of

$$\rho_s = \frac{316 \times 10^{-12}}{10 \times 10^{-4}}$$

$$= 316 \, nC \, m^{-2}$$

Each unit charge will radiate a unit of flux, and so the flux density in the air gap will be

$$D = 316 \, nC \, m^{-2}$$

This flux density will remain constant throughout the capacitor. With this flux density, the electric field strength in each air gap is 316×10^{-9}

$$E_a = \frac{316 \times 10^{-9}}{8.854 \times 10^{-12}}$$

$$= 35.7 \, kV \, m^{-1}$$

Let us now turn our attention to the dielectric. We have a flux density of 316 nC m^{-2} at the surface of the capacitor. However, we have just seen that (Equation (6.8))

$$D = \varepsilon_0 E + P$$

or

$$D = \varepsilon_0 \varepsilon_r E$$

So, the E field in the dielectric is

$$E = \frac{D}{\varepsilon_0 \varepsilon_r}$$

$$= \frac{316 \times 10^{-9}}{8.854 \times 10^{-12} \times 2.5}$$

$$= 14.3 \, kV \, m^{-1}$$

The difference between the air and dielectric field strengths is due to the bound charges on the surface of the dielectric. So, as $D = \varepsilon_0 E$, the surface charge density (or flux density) is

$$\rho_d = \varepsilon_0 \times (14.3 - 35.7) \times 10^3$$

$$= -8.854 \times 10^{-12} \times 21.4 \times 10^3$$

$$= -190 \, nC \, m^{-2}$$

Figure 6.3c shows the distribution of flux density and surface charges throughout the capacitor.

At first sight, the work we have done in this section may not appear to be of much practical benefit. After all, we are quite happy to work with the relative permittivity of a material, so why should we concern ourselves with polarization? The answer to this question should become clear in the next section when we examine what happens to an electric field as it crosses a dielectric boundary.

6.3   BOUNDARY RELATIONSHIPS

In the example at the end of the last section, we encountered a composite dielectric capacitor. This meant that the electric field in the capacitor had to cross a boundary between two dielectrics. The question of what happens to the E field is very important – light and radio waves are electromagnetic in form, and it would be nice to know what happens to these signals as they pass through a dielectric.

Figure 6.5a shows an electric field crossing the boundary between two dielectrics. To analyze this situation we will split the incident and transmitted field into their tangential and perpendicular components.

Figure 6.5b shows the tangential component of the E field either side of the boundary. Let us consider a rectangular path ABCD that lies either side of the boundary. Now, the work done in moving a unit point charge from point A around the path and back again must equal zero. (We have moved the charge around a closed loop and so have not gained or lost potential.) So,

$$\int_{ABCDA}^{0} E \, dl = E_{t1}dl + E_{n1}dw - E_{t2}dl - E_{n2}dw \tag{6.11}$$

$$= 0$$

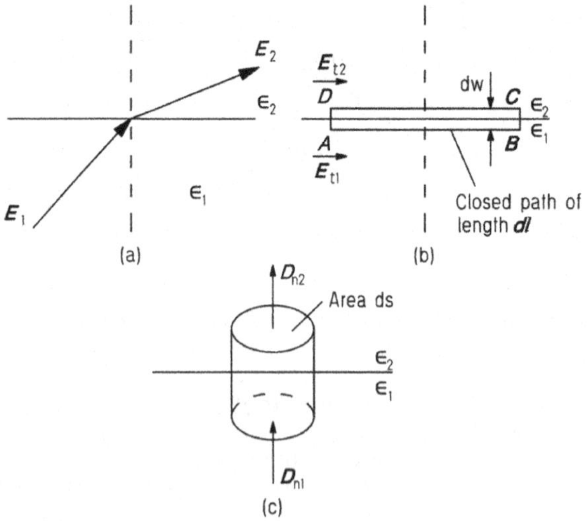

FIGURE 6.5   (a) An electric field crossing the boundary between two dielectrics, (b) tangential E field at the boundary and (c) normal component of D field at the boundary.

If we make this path extremely thin, i.e., $dw \to 0$, we get

$$E_{t1}dl - E_{t2}dl = 0$$

and so,

$$E_{t1}dl = E_{t2}dl \qquad (6.12)$$

i.e., the tangential component of the E field is continuous across the boundary.

To find out what happens to the normal component of the fields, let us construct a pill-box with the top face in medium 2, and lower face in medium 1, as shown in Figure 6.5c. Now, electric flux entering the pill-box does so via the lower face and flux leaving the pill-box does so via the upper face. If there are no surface charges, the flux entering must equal the flux leaving, i.e.,

$$\psi_1 = \psi_2 \qquad (6.13)$$

or

$$D_{n1}ds = D_{n2}ds$$

and so,

$$D_{n1} = D_{n2} \qquad (6.14)$$

i.e., the normal component of the D field is continuous across the boundary if there are no bound charges.

As we have seen in the previous section, polarization causes bound charges to appear at the surface between the two dielectrics. These charges will radiate additional flux out of the pill-box, and so Equation (6.13) becomes

$$\psi_2 = \psi_1 + q_s \qquad (6.15)$$

where q_s, is the surface charge. If we use the surface charge density, we can write

$$D_{n2}ds = D_{n1}ds + \rho_s ds$$

and so,

$$D_{n2} = D_{n1} + \rho_s \qquad (6.16)$$

i.e., the normal component of the D field is discontinuous across the boundary if there are bound charges. The amount of discontinuity is equal to the surface charge density.

Before we consider an example, let us digress slightly and examine what happens at the boundary between a conductor and air. Let us introduce some charges into the conductor. These charges will repel each other until they end up in equilibrium on the surface of the conductor. Thus, we will have a surface charge density similar to that which we met in the last example. The fact that charges are in equilibrium on the surface of the conductor means that the tangential component of the E field must be zero inside the wire.

As regards the surface charges, these will radiate flux in a direction normal to the surface. (If there were a tangential component to the flux, this would generate a tangential E field and that is not allowed!) So, we can say

$$E_t = 0 \tag{6.17}$$

and

$$E_t = \frac{\rho_s}{\varepsilon_0} \tag{6.18}$$

Exactly the same result applies if the conductor is placed in an E field. This is of some practical benefit – if the conductor is an earthed plate, the surface charge will dissipate to ground and so the field will not pass through. This is the principle behind the earthed shield in coaxial cable.

Example 6.2

An E field makes an angle of 30° to the horizontal at the boundary between air and a dielectric. If the dielectric has $\varepsilon_r = 6$, determine the angle of refraction. Assume that there are no bound charges.

Solution

The tangential component of the E field is continuous across the boundary. So, the tangential E field in the dielectric is

$$E_t = E\cos 30°$$

$$= \frac{2}{\sqrt{3}}E$$

As there are no bound charges at the surface, the normal component of the D field is continuous across the boundary. Thus,

$$D_n = D\sin 30°$$

or

$$D_n = \varepsilon_0 E \sin 30°$$

Now, the dielectric has $\varepsilon_r = 6$ and so,

$$6 \times \varepsilon_0 \times E_a = \varepsilon_0 E \sin 30°$$

or

$$E_n = \frac{E}{6}\frac{1}{\sqrt{3}}$$

We can find the angle of refraction from

$$\tan\alpha = \frac{E_n}{E_t}$$

$$= \frac{1}{6\sqrt{3}} \times \frac{\sqrt{3}}{2}$$

$$= \frac{1}{12}$$

Thus, the refracted E field makes an angle of 4.8° to the horizontal. We should expect this because the normal component of the field has been significantly reduced, whereas the tangential component has hardly altered.

6.4 DIELECTRIC STRENGTH AND MATERIALS

Although this is the last section in this chapter, it is probably the most important from an engineering point of view. As we have already seen, an electric field tends to polarize the atoms in a dielectric. If the field is great enough, the dielectric could break down by direct ionization of the atoms. This causes the leakage current to rise dramatically, resulting in a very rapid rise in temperature and an explosion! The dielectric strength of an insulator is the maximum field strength that the material can stand before ionization occurs. Table 6.1 lists the relative permittivities and dielectric strengths of some common insulating materials.

A glance at this table shows a wide spread in the maximum field strength that insulators can withstand. However, they are all of the order of several million volts per metre. It may be thought that such a high E field will not occur in practice.

TABLE 6.1
Relative Permittivity and Dielectric Strength for a Variety of Dielectric Materials

Material	e_r	Dielectric Strength (V m^{-1})
Air (atmospheric pressure)	1.0	3×10^6
Mineral oil	2.3	15×10^6
Rubber	2.3–4.0	25×10^6
Paper	2.0–4.0	15×10^6
Polystyrene	2.6	20×10^6
Glass	4.0–10	30×10^6
Mica	6.0	200×10^6

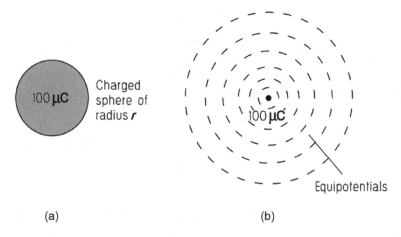

FIGURE 6.6 (a) A charge sphere and (b) the equipotential plot.

However, as we have already seen, the field inside a capacitor does approach the maximum listed in the table.

So, we must use care when designing a capacitor not to produce too large a field strength. What about the shape of the conductors? Is there one particular shape that results in the lowest field strength? To answer these questions let us consider a charged sphere and calculate the field at the surface.

Figure 6.6 shows a sphere that has a total charge of 100 μC on the surface. We can use Gauss' law to replace this sphere by a point charge of magnitude 100 μC at the centre of the sphere. Thus, the E field at the surface is simply

$$E_r = \frac{100 \times 10^{-6}}{4\pi\varepsilon_0 r^2}$$

$$= \frac{9 \times 10^5}{r^2}$$

(6.19)

where r is the radius of the sphere.

So, if we halve the radius of the sphere, the surface E field increases by a factor of 4. Clearly there will come a point where the E field will exceed the dielectric strength of air. Taking a maximum E field of $3 \times 10^6 \,\mathrm{V\,m^{-1}}$ gives a minimum radius of 55 cm. If we make the sphere any smaller, we could ionize the surrounding air, resulting in corona discharge of the sphere. (The fact that small conductors generate high field strength is put to good use in lightning conductors. These have a pointed end that produces a large electric field. When the field is great enough, a corona discharge is created (the lightning strike) and charge flows along the lightning conductor.)

To guard against ionization in a capacitor, we must ensure that the capacitor plates are as uniform as possible – any points on the plates will generate a large electric field. So, the metal plates used in capacitors must be flat.

Two types of dielectric are currently used in capacitors: non-polar materials and polar materials. Non-polar materials are those listed in Table 6.1. Such dielectrics are used to give capacitors of value up to 0.1 μF.

Large-value capacitors tend to use polar materials, or electrolytes. These materials conduct electricity, and so why use them in capacitors? The answer lies in the construction of polarized capacitors. One of the capacitor plates has a very thin layer of aluminium oxide deposited on it. A conducting electrolyte separates this oxide from the other plate, and so the dielectric thickness is simply the thickness of the oxide. Thus, the capacitance of this arrangement can be high.

Anyone who has used electrolytic capacitors will know that they have to be connected the right way round. This is because the electrolyte will conduct in one particular direction. Thus, if we incorrectly connect an electrolytic capacitor, a very large current flows, resulting in a large bang and clouds of vaporized electrolyte!

7 Ferromagnetic Materials and Components

In this chapter, we will re-examine transformers and the effect of a magnetic field on ferrous materials. As we saw in Section 3.11, ferrous cores cause the inductance of a coil to increase. Thus, iron-cored inductors can be physically smaller than their air-cored counterparts. One effect of an iron core is that it provides a low reluctance path to magnetic flux. This reduces any flux leakage and leads to a more efficient transformer.

In common with dielectrics, a magnetic field produces a magnetic dipole moment in the ferrous material and this is where we begin our studies.

7.1 MAGNETIC DIPOLES AND PERMANENT MAGNETS

In the last chapter, we developed a model of dielectric polarization in which an external field distorted the atoms into electric dipoles. In magnetic materials, we can use a similar model in which the material has a random distribution of tiny permanent magnets, or magnetic dipoles (see Figure 7.1a).

Now, when we apply an external magnetic field to the material, some of the dipoles will line up in the direction of the field (Figure 7.1b). If we continue to increase the external field, there comes a point when all the dipoles line up (Figure 7.1c).

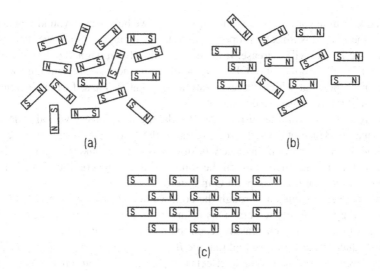

FIGURE 7.1 (a) Randomly distributed magnetic dipoles, (b) distribution of magnetic dipoles due to external field and (c) saturation of magnetic dipoles.

This is known as saturation. If we now remove the external field, some of the magnetic dipoles will remain lined up in their new positions, so creating a permanent magnet.

What happens if we introduce a unit pole into some magnetic material? This pole will tend to re-align some of the magnetic dipoles surrounding it. If we then introduce another pole at a certain distance from the original, this new pole will experience a force from the surrounding dipoles and not from the original monopole. Thus, the material reduces the original force field. We can check the validity of this by noting that the magnetic force field in air is given by

$$H = \frac{B}{\mu_0} \tag{7.1}$$

If the pole is in a magnetic medium, the force field is

$$H = \frac{B}{\mu_0 \mu_r} \tag{7.2}$$

which is lower for air. This implies that it is the polarization of the material that causes a permeability greater than air.

7.2 POLARIZATION AND THE *B/H* CURVE

When we discussed dielectrics in the previous chapter, we introduced the idea of dielectric polarization. As we have just seen, we have a similar situation in magnetic materials and so we can introduce a magnetic polarization vector, *M*. Thus, the flux density can be modified to

$$B = \mu_0 H + M \tag{7.3}$$

Let us take a moment to examine this equation. If we have a coil wound on a ferrous former, the *H* field is directly proportional to the coil current. So, if we increase the current, the *B* field will increase. If the core is initially unmagnetized, a plot of *B/H* (Figure 7.2) will follow curve OP. If we continue to increase the coil current, there will come a point at which the core saturates, point P. Thus, further increases in current will not increase the flux density.

Let us now decrease the current. As *H* falls, the flux density will also decrease. Unfortunately, *M* is not directly proportional to the field – this is due to some of the magnetic dipoles staying in their new positions. So, if we make *H* equal to zero, by removing the current, there will still be some residual magnetic flux. This is known as the remanence of the core, point Q in Figure 7.2.

If we reverse the current through the coil, we need to produce an *H* field of OR to reduce the flux density to zero. This is known as the coercivity of the core. If we continue to increase the current we reach the reverse saturation point *S*. If we then start to reduce the current, we find that the *B/H* curve follows a new path, STUP. So we can never remove the magnetic effects completely. (We could heat the core up to the Curie temperature, at which point it does lose its remanence. However, this would cause the insulation surrounding the wire to melt, so destroying the coil!)

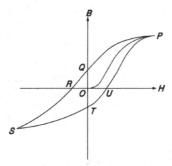

FIGURE 7.2 *B/H* curve of a magnetic sample.

The fact that the *B/H* curve is not linear means that M is not directly proportional to B. Thus, we cannot use the equivalent of electric susceptibility. Instead, we have to accept that the relative permeability of the core is non-linear and must be found from the *B/H* curve.

7.3 BOUNDARY RELATIONSHIPS

As stated in Section 6.3, light is an electromagnetic wave, i.e., it has an electric and a magnetic field. In that section, we were interested in what happened at the boundary between two dielectrics. Here, we will concern ourselves with the effect of a change in magnetic media on the H field.

Figure 7.3a shows a magnetic field crossing the boundary between two ferrous materials. To analyze this situation, we will split the incident and transmitted fields into their tangential and perpendicular components.

Figure 7.3b shows the tangential component of the H field either side of the boundary. Let us consider a rectangular path ABCD that lies either side of the boundary. Now, Ampère's law states that the line integral of the H field around a closed path must equal the enclosed current. If we have no surface currents, we get

$$\int\limits_{ABCDA}^{0} H \ \mathrm{d}l = H_{t1}\mathrm{d}l + H_{n1}\mathrm{d}w - H_{t2}\mathrm{d}l - H_{n2}\mathrm{d}w$$

$$= 0$$

(7.4)

If we make this path extremely thin, i.e., $\mathrm{d}w \to 0$, we get

$$H_{t1}\mathrm{d}l - H_{t2}\mathrm{d}l = 0$$

and so,

$$H_{t1} = H_{t2}$$

(7.5)

i.e., the tangential component of the H field is continuous across the boundary provided there are no surface currents.

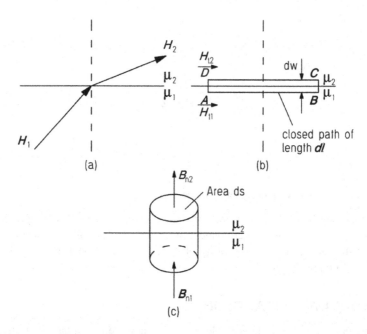

FIGURE 7.3 (a) A magnetic field crossing the boundary between two different ferrous materials, (b) tangential H field at the boundary and (c) normal component of B field at the boundary.

It is interesting to note that if surface currents are set up in the new media, the transmitted H field will be severely attenuated and may even be zero for a good conductor. These surface currents are known as eddy currents. Coaxial cable uses this property to induce eddy current in the shielding. As the shielding is grounded, the eddy currents will be conducted to earth and lost.

To find out what happens to the normal component of the field, let us construct a pill-box with the top face in medium 2 and lower face in medium 1 as shown in Figure 7.3c. Now, magnetic flux entering the pill-box does so via the lower face and flux leaving the pill-box does so via the upper face. As there can be no monopoles on the surface, the flux entering must equal the flux leaving, i.e.,

$$\phi_1 = \phi_2 \tag{7.6}$$

or

$$B_{n1}ds = B_{n2}ds$$

and so,

$$B_{n1} = B_{n2} \tag{7.7}$$

i.e., the normal component of the B field is continuous across the boundary.

7.4 IRON-CORED TRANSFORMERS

Iron-cored transformers use the fact that the flux density in ferrous material is greater than in air for the same H field. Such devices are found in mains powered appliances and electrical substations. Figure 7.4 shows the general form of a transformer.

As can be seen from this figure, we have two coils wound on a former. Let us call the left-hand coil the primary and the right-hand coil the secondary. Now, convention dictates that the primary is connected to the supply and the secondary is connected to the load. As we have previously seen, an alternating voltage causes an alternating flux in the core. As the core has a permeability greater than that of air, the reluctance of the core is less than air; see Equation (3.38). So, most of the flux will be concentrated in the core. This alternating flux flows through the secondary coil where it generates an alternating voltage at the secondary terminals.

As we saw in Chapter 3, the primary voltage is given by (Equation (3.43))

$$V_1 = N_1 \frac{d\phi}{dt}$$

and so,

$$\frac{d\phi}{dt} = \frac{V_1}{N_1} \tag{7.8}$$

This alternating flux generates a voltage at the secondary terminals of

$$V_2 = N_2 \frac{d\phi}{dt}$$

$$= N_2 \frac{V_1}{N_1}$$

i.e.,

$$\frac{V_2}{V_1} = \frac{N_1}{N_2} \tag{7.9}$$

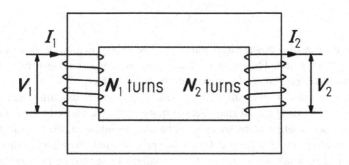

FIGURE 7.4 Schematic of a typical transformer.

If N_1, is greater than N_2, the secondary voltage is less than the primary, i.e., we have a step-down transformer. If N_1, is less than N_2, the secondary voltage is greater than the primary and we have a step-up transformer. So, the turns ratio determines the secondary voltage, but what happens to the current?

Well, let us suppose that the transformer is ideal, i.e., there are no losses in the windings, and that the core has a linear *B/H* characteristic. Power taken by the secondary load has to be supplied by the primary and so

$$V_1 I_1 = V_2 I_2$$

By substituting for V, from Equation (7.9), we get

$$V_1 I_1 = V_1 \frac{N_2}{N_1} I_2$$

i.e.,

$$\frac{I_2}{I_1} = \frac{N_1}{N_2} \qquad (7.10)$$

Thus, if we are considering a step-down transformer, the voltage will decrease by the turns ratio, but the current will increase by the same amount. The example at the end of this section shows this best.

Let us return to Equations (7.9) and (7.10). After some rearranging, we get

$$V_2 = \frac{N_2}{N_1} V_1 \qquad (7.11)$$

and

$$I_2 = \frac{N_2}{N_1} I_1 \qquad (7.12)$$

If we divide Equation (7.11) by Equation (7.12) we get, after some rearranging,

$$R = \left(\frac{N_1}{N_2}\right)^2 R_2 \qquad (7.13)$$

So, if we have a step-down transformer, the secondary load appears to be greater from the primary side. This property means that we can use transformers to match source and load impedances for maximum power transfer.

Before we consider an example, let us take a moment to examine the losses that occur in an iron-cored transformer. As the transformer uses wire in the primary and secondary coils, there will be some resistive loss. In addition, the very fact that we are using coils means that there will be some primary and secondary inductance. As Equation (7.13) shows, we can refer the secondary losses to the primary as shown in Figure 7.5. We can measure these losses, from the primary side, by performing

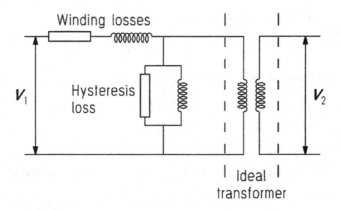

FIGURE 7.5 Basic equivalent circuit of a transformer.

a short-circuit test. We must not do these tests at the rated voltage because the high currents will destroy the transformer.

So, we can easily account for the winding losses. However, what about the shape of the B/H curve? Will this introduce additional loss? The answer to this question is yes. The B/H curve introduces a loss known as hysteresis loss. Let us refer to the B/H curve shown in Figure 7.2. If the supply voltage is an alternating source, the current and H field will also alternate. This means that we go once round the B/H curve for every cycle of the supply current. We can find the energy stored in the magnetic field from

$$\text{energy} = \int H \, dB$$

which is simply the area of the B/H curve. So, we find the hysteresis loss per cycle, and hence the power loss, if we know the area enclosed by the curve. If the area enclosed by the curve is small, the transformer is very efficient.

Example 7.1

An ideal transformer is wound on an iron former with 200 turns on the primary and 30 turns on the secondary. Determine the secondary voltage if the primary is connected to a 240 V source. If a secondary load takes 5 A, determine the primary current and the equivalent load on the primary.

Solution

We have an ideal transformer with 200 turns on the primary and 30 on the secondary. So, from Equation (7.9), we have

$$V_2 = \frac{N_2}{N_1} V_1$$

$$= \frac{30}{200} 240 = 36 \, \text{V}$$

If the secondary takes 5 A, the secondary power will be

$$P_2 = 36 \times 5 = 180\,\text{W}$$

This power must come from the primary, and so

$$180 = V_1 \times I_1$$

$$= 240 \times I_1$$

Thus, the primary current is 0.75 A.

Now, the secondary load takes 5 A from a 36 V supply. So, the secondary load is

$$R_2 = \frac{36}{5} = 7.2\ \Omega$$

The primary voltage is 240 V, and the primary current is 0.75 A. Thus, the primary load is

$$R_1 = \frac{240}{0.75} = 320\ \Omega$$

So, the 7.2 Ω secondary load appears to be a 320 Ω load when viewed from the primary.

7.5   ELECTRICAL MACHINERY

As we saw in Chapter 3, a force is exerted on a current-carrying wire placed in a magnetic field. So, what happens if we wind some turns on to a former that is free to rotate in a magnetic field?

Figure 7.6a shows a bobbin that is free to rotate on its axis. There are a number of turns on this former, which is placed between the poles of a permanent magnet. When we energize the coil, the coil current produces a magnetic field. This interacts with the field from the permanent magnet, so generating torque on the bobbin. This is known as the motor effect and it is used in moving-coil meters.

The force on one conductor is given by

$$F = BIl\ \sin\theta \tag{7.14}$$

where B is the flux density in the air gap, I is the current through the coil, l is the length of the bobbin and θ is the angle between a line drawn normal to the axis of the coil and the field – see Figure 7.6. (We came across a form of Equation (7.14) in Chapter 3 when we considered the Biot-Savart law. Yet again the angle θ is used so Equation (7.14) can be expressed using the vector cross product.)

Let us examine the factor $\sin\theta$ a bit more closely. If the coil is vertical (Figure 7.6b), θ is 0° and, according to Equation (7.14), the rotational force is zero. If we draw the field due to the current in the wire, we can see that both conductors experience a force acting into the coil. Thus, there is no rotational force on the wire.

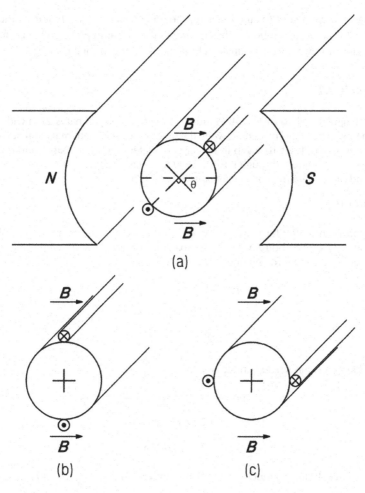

FIGURE 7.6 (a) A basic electric motor, (b) position for minimum torque and (c) position for maximum torque.

If the coil is horizontal (Figure 7.6c), θ is 90° and the rotational force should be a maximum. If we draw in the field generated by the current in the wire, we can see that the field lines on the left-hand conductor indicate an upward force and those on the right-hand conductor show a downward force. So each conductor will experience maximum force.

We can turn this motor into a generator by driving the bobbin by some means. If we do this, the conductors will cut the flux from the permanent magnet, so inducing a voltage in the conductors. Specifically, the voltage induced in one conductor is

$$v(t) = B \times ld \times \omega \times \sin \omega t \qquad (7.15)$$

where ld is the area of the coil, ω is the angular velocity of the bobbin and the $\sin \omega t$ is there to account for the rotation of the coil. If we can extract this voltage by using slip rings, we can use it to supply current to an external load. Obviously, the prime mover that is rotating the bobbin has to supply this drain in power. This is similar to the transfer of power from a transformer primary to the secondary.

Example 7.2

A rectangular coil consists of 40½ turns wound on a former that is 2 cm wide and 5 cm long. The coil is placed between the poles of a permanent magnet that generates a flux density of 10 mWb in the air gap. Determine the maximum torque on the coil if the current through it is 5 mA. If the coil is rotated at 100 revolutions per second, determine the voltage generated.

Solution

The former has 40½ turns wound on it, i.e., there are 40 wires on the top of the coil and 41 on the bottom. Now, the maximum torque occurs when the coil is horizontal. So, the force on one conductor is

$$F = BIl \sin \theta$$

$$= 10 \times 10^{-3} \times 5 \times 10^{-3} \times 5 \times 10^{-2}$$

$$= 2.5 \ \mu N$$

The torque on this single conductor is

$$torque = force \times radius$$

$$= 2.5 \times 10^{-6} \times 1 \times 10^{-2}$$

$$= 25 \times 10^{-9} \ N \ m$$

As we have 40 top conductors and 41 bottom conductors, the total torque is

$$torque = (40 + 41) \times 25 \times 10^{-9}$$

$$= 2 \times 10^{-6} \ N \ m$$

If the motor is turned into a generator rotating at 100 revolutions per second, the voltage generated will be (Equation (7.15))

$$v = B \times ld \times \omega$$

$$= 10 \times 10^{-3} \times 2 \times 10^{-2} \times 5 \times 10^{-2} \times 2\pi \times 100$$

$$= 6.3 \ mV \ per \ conductor$$

With 81 conductors, the terminal voltage will be

$$v = 8.1 \times 6.3 \times 10^{-3}$$

$$= 0.5 \ V \ at \ a \ frequency \ of \ 100 \ Hz.$$

So, the generator is not particularly good, −0.5 V will not do much. If we want to increase the voltage, we have to increase the dimensions of the generator and/or increase the flux density in the air gap.

7.6 THE MAGNETIC CIRCUIT

So far we have only considered simple magnetic structures. This enabled us to study the basics of magnetic fields. However, there are many complicated magnetic structures in use today. This is where we can make good use of an electrical analogy. The analysis of a magnetic circuit is best done by an example.

Figure 7.7a shows the cross section through a relay that has energizing coils on the left-hand and centre limbs. The right-hand limb has a magnetic lever in an air gap that is connected to some external contacts. When we energize the left-hand coil, the relay contacts close, so energizing the centre-limb coil. This generates sufficient flux to keep the contacts closed regardless of the current in the left-hand limb. We need a force of 0.5 N to close the contacts, and the contact magnet produces a flux of 10 μWb. The relative permeability of the core is constant and has a value of 100. We require to find the current needed in either coil to cause the contacts to close.

To analyze this situation, we will use an electrical analogy. Each coil is a source of magnetic potential and so we can represent them by voltage sources. We can represent flux by current and core reluctance by resistance. Thus, we can generate the electrical analogy shown in Figure 7.7b.

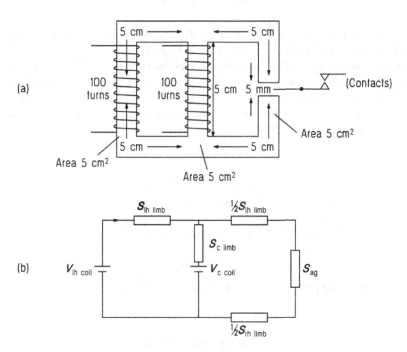

FIGURE 7.7 (a) A two-coil relay actuator and (b) electrical analogy for the two-coil relay actuator.

Now, we do not know the flux needed to close the contacts and so this must be our starting point. From Section 3.11 we know that the force between two magnetic surfaces is (Equation (3.69))

$$\text{force} = \frac{1}{2}\frac{B^2}{\mu_0} \times \text{area}$$

where B is the flux density in the air gap. We require a force of $0.5\,\text{N}$, and so we have

$$0.5 = \frac{1}{2}\frac{B^2}{4\pi \times 10^{-7}} 5 \times 10^{-4}$$

giving

$$B^2 = 2.5 \times 10^{-3}$$

So,

$$B = 5 \times 10^{-2}\ \text{Wb m}^{-2}$$

Now, the area of the face generating this flux density is $5\,\text{cm}^2$, and so the total flux in the air gap is

$$\phi = 25 \times 10^{-6}\ \text{Wb}$$

As the flux from the magnet is $10\ \mu\text{Wb}$, the flux needed from the magnetic circuit is $15\ \mu\text{Wb}$. So, the flux in the right-hand limb is $15\ \mu\text{Wb}$.

We now need to determine the reluctance of the right-hand limb. As Figure 7.7b shows, there are three components to the total reluctance: the upper half of the limb, the air gap and the lower half of the limb. The length of this limb is $10\,\text{cm}$, in which there is a 5-mm air gap. As the air gap is very small compared with the length of the limb, we can write

$$S_{\text{rh limb}} = S_{\text{limb}} \times S_{\text{air}}$$

$$\frac{10 \times 10^{-2}}{100_{\mu 0} \times 5 \times 10^{-4}} + \frac{5 \times 10^{-3}}{0_{\mu 0} \times 5 \times 10^{-4}}$$

$$= 1.6 \times 10^6 + 8 \times 10^6$$

$$= 9.6 \times 10^6\ \text{At Wb m}^{-1}$$

Thus, the magnetic potential across the right-hand limb is

$$V_{\text{rh limb}} = S_{\text{rh limb}} \times \phi_{\text{rh limb}}$$

$$= 9.6 \times 10^6 \times 15 \times 10^{-6}$$

$$= 143\ \text{At}$$

This must be the potential across the centre limb, which is made up of the potential generated by the centre coil and the reluctance of the centre limb. If we assume that the centre coil is not energized, this circuit only consists of the reluctance given by

$$S_{c \, limb} = \frac{5 \times 10^{-2}}{100_{\mu 0} \times 5 \times 10^{-4}}$$

$$= 8 \times 10^5 \ At \ Wb^{-1}$$

Thus, the flux down the centre limb is

$$\phi_{c \, limb} = \frac{V_{c \, limb}}{S_{c \, limb}}$$

$$= \frac{143}{8 \times 10^5}$$

$$= 1.8 \times 10^{-4} \ Wb$$

This must be added to the flux down the right-hand limb, and so the flux from the left-hand limb is

$$\phi_{lh \, limb} = \phi_{c \, limb} + \phi_{rh \, limb}$$

$$= 1.8 \times 10^{-4} + 15 \times 10^{-6}$$

$$= 1.95 \times 10^{-4} \ Wb$$

This flux has to flow through the reluctance of the left-hand limb, which is

$$S_{lh \, limb} = \frac{10 \times 10^{-2}}{100_{\mu 0} \times 5 \times 10^{-4}}$$

$$= 1.6 \times 10^5 \ At \ Wb^{-1}$$

Hence, the magnetic potential is

$$V_{lh \, limb} = \phi_{lh \, limb} + S_{lh \, limb}$$

$$= 1.95 \times 10^{-4} \times 1.6 \times 10^6$$

$$= 312 \ At$$

This must be added to the potential across the centre/right-hand limb. So, the potential generated by the left-hand coil is

$$V_{lh \, coil} = V_{lh \, limb} + V_{rh \, limb}$$

$$= 312 + 143$$

$$= 455 \ At$$

As there are 100 turns on the left-hand coil, the current has to be

$$I_{\text{lh coil}} = \frac{455}{100}$$

$$= 4.55 \text{ A}$$

Now, if no current flows through the coil on the left-hand limb, the centre limb has to supply the flux. As we have already seen, the flux down the right-hand limb is 15 µWb, and this must be added to the flux down the left-hand limb given by

$$\phi_{\text{lh limb}} = \frac{V_{\text{lh limb}}}{S_{\text{lh limb}}}$$

$$= \frac{143}{1.6 \times 10^6}$$

$$= 8.94 \times 10^{-5} \text{ Wb}$$

Thus, the total centre-limb flux is

$$\phi_{\text{c limb}} = \phi_{\text{lh limb}} + \phi_{\text{rh limb}}$$

$$= 8.94 \times 10^{-5} + 15 \times 10^{-6}$$

$$= 1 \times 10^{-4} \text{ Wb}$$

And so the magnetic potential across the centre limb must be

$$V_{\text{c limb}} = \phi_{\text{c limb}} + S_{\text{c limb}}$$

$$= 1 \times 10^{-4} \times 8 \times 10^{-5}$$

$$= 80 \text{ At}$$

Therefore, the coil on the centre limb has to supply

$$V_{\text{c coil}} = V_{\text{c limb}} + V_{\text{lh limb}}$$

$$= 80 + 143$$

$$= 223 \text{ At}$$

As the centre limb has 100 turns wound on it, the centre limb is

$$I_{\text{c limb}} = \frac{223}{100}$$

$$= 2.2 \text{ A}$$

This example has shown how to analyze a magnetic circuit by using an electrical analogy. We were able to use the reluctance of the core because we assumed that the core had a linear B/H curve. If we cannot assume this, we would have to calculate B, find H from the curve, multiply by the length of the circuit, find the new flux and so on. This is the method that must be used in Problem 7.3.

8 Waves in Transmission Lines

When we turn on a light, the light will not turn on instantaneously. Instead, it takes a finite time to turn on. (Although it appears to us that the light turns on immediately, the voltage from the switch to the light takes a small amount of time to travel from the switch to the light.) It is this travelling wave that concerns us here. We will consider a transmission line and analyze what happens to the voltage and current as the disturbance propagates.

8.1 THE LOSSLESS TRANSMISSION LINE

As we have seen in Chapters 2–4, a transmission line has internal resistance, external conductance, capacitance and inductance. We could analyze a transmission line using the full model, but it is informative to use a section of the *lossless* line in which there is zero resistance and conductance as shown in Figure 8.1.

Figure 8.1 shows a small section of the cable of length Δx with the lumped parameters, L and C being used. The inductor gives a back emf by virtue of the changing current and the capacitor takes a current by virtue of the changing voltage. Applying Kirchhoff's laws gives

$$v(x, t) - L\Delta x \frac{dI(x, t)}{dt} - \left(v(x, t) + \frac{dv(x, t)}{dx} \Delta x \right) = 0 \tag{8.1}$$

$$i(x, t) - C\Delta x \frac{dV(x + \Delta x, t)}{dt} - \left(i(x, t) + \frac{di(x, t)}{dx} \Delta x \right) = 0 \tag{8.2}$$

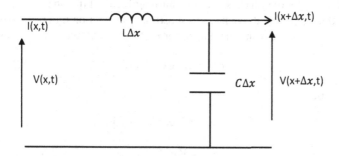

FIGURE 8.1   Section of a lossless transmission line.

By letting $\Delta x \to dx$, (8.1) and (8.2) become

$$-\frac{dv(x, t)}{dx} = L\frac{di(x,t)}{dt} \qquad (8.3)$$

$$-\frac{di(x, t)}{dx} = C\frac{dv(x,t)}{dt} \qquad (8.4)$$

These equations are linked – (8.3) has the differential of v wrt distance, while (8.4) has the differential of v wrt time. If we differentiate (8.3) with respect to x and (8.4) with respect to t, we get

$$\frac{d^2v(x, t)}{dx^2} = -L\frac{d^2i(x,t)}{dxdt} \qquad (8.5)$$

$$\frac{d^2i(x, t)}{dxdt} = -C\frac{d^2v(x,t)}{dt^2} \qquad (8.6)$$

Substitution of (8.6) into (8.5) yields

$$\frac{d^2v(x, t)}{dx^2} = LC\frac{d^2v(x,t)}{dt^2} \qquad (8.7)$$

A similar derivation for the current gives

$$\frac{d^2i(x, t)}{dx^2} = LC\frac{d^2i(x,t)}{dt^2} \qquad (8.8)$$

Equations (8.7) and (8.8) are known as the wave equations and possible solutions are that of a travelling wave in the positive or negative direction, i.e.,

$$v(x, t) = V_o\cos(\omega t \pm \beta x) \qquad (8.9)$$

This equation shows that the voltage (and current) varies with time and distance. If we take the negative in (8.9), we have a positive travelling wave as we will see shortly. The parameter β is the phase shift per unit length called the phase coefficient with units of rad m^{-1}. To find it, we need to substitute the solution into the wave equation (8.8). This is best done using the phasor form of (8.9) and Euler's expression:

$$e^{j\varnothing} = \cos\varnothing + j\sin\varnothing$$

and so (8.9) becomes

$$v(x, t) = V_o\cos(\omega t - \beta x)$$

$$= V_o e^{j(\omega t - \beta x)}$$

$$= V_o e^{j\omega t}e^{-j\beta x}$$

The advantage of this method is that differentiation is very simple.

$$\frac{dv(x, t)}{dt} = j\omega V_o e^{j\omega t} e^{-j\beta x} = j\omega v(x, t)$$

$$\frac{d^2 v(x, t)}{dt^2} = (j\omega)^2 V_o e^{j\omega t} e^{-j\beta x} = (j\omega)^2 v(x, t)$$

So, (8.8) becomes

$$(-j\beta)^2 v(x, t) = LC(-j\omega)^2 v(x, t)$$

Hence,

$$\beta = \omega\sqrt{LC} \tag{8.10}$$

Equation (8.9) is only a solution to the wave equation if β takes on this value. If we use our definitions of L (external only) and C from (3.52) and (2.37), we find that the result does not depend on the dimensions of the cable

$$\beta = \omega\sqrt{\mu\varepsilon} = \frac{\omega}{c} \tag{8.11}$$

where c is the speed of light *in the surrounding dielectric*. There is a clear link between a circuit point of view and that of field theory. We will see this again in the next section.

Figure 8.2 shows a plot of (8.9) at time $t = 0$ and sometime later at $t = t_1$. Let us consider the movement of the constant phase point A that starts at $x = 0$ and $t = 0$. At time $t = t_1$, this point has moved to $x = x_1$ and so the velocity is x_1/t_1. This constant phase point has an amplitude of V volt. For this to occur, the cosine term in (8.9) must be unity, i.e.,

$$v(0,0) = V_o \cos(0) = v(x_1, t_1) = V_o \cos(\omega t_1 - \beta x_1)$$

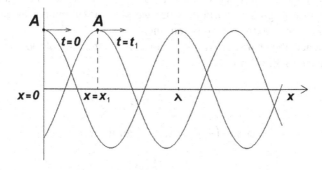

FIGURE 8.2   A travelling wave.

This implies that $\omega t_1 = \beta x_1$ and so the phase velocity is

$$v_p = \frac{\omega}{\beta} = \frac{1}{\sqrt{LC}} = \frac{1}{\sqrt{\mu\varepsilon}} \tag{8.12}$$

This is the speed of light in the surrounding medium. If it is air, $\mu = 4\pi \times 10^{-7}$ H m^{-1} and $\varepsilon = 8.854 \times 10^{-12}$ F m^{-1} and so $v_p = 3 \times 10^8$ m s^{-1} – the speed of light in air. If we have a denser, non-magnetic dielectric, the permeability is still that of free-space but the permittivity is increased by ε_r and so the speed of light is lower. (This is the situation in transmission lines and glass.)

Before we go on to discuss impedance, let us take another look at Figure 8.2. The wave at time $t = 0$ repeats itself every full cycle of distance. This distance is known as the wavelength, λ. So,

$$v(0,0) = V_o \cos(0) = v(\lambda,0)$$

This can be valid if the cosine term in (8.9) is unity. This occurs at cos 0 or cos 2π. Taking the second gives

$$\beta\lambda = 2\pi$$

and so

$$\lambda = \frac{2\pi}{\beta}$$

$$= \frac{1}{f\sqrt{LC}}$$

$$= \frac{c}{f} \tag{8.13}$$

Let us now turn to impedance, but not the kind we can measure using a resistance meter. What we are concerned with is the characteristic impedance which is the impedance seen by a travelling signal – a surge impedance or the characteristic impedance. This is of great importance when we want maximum power transfer; we must match the source to the transmission line.

Like resistance, the characteristic impedance, Z_o, is the ratio of the voltage to the current. By using (8.3), we get

$$-\frac{dv(x,t)}{dx} = L\frac{di(x,t)}{dt}$$

and so $-(-j\beta)v(x,t) = j\omega Li(x,t)$

$$\frac{v(x,t)}{i(x,t)} = \frac{\omega L}{\beta}$$

$$\text{Therefore, } Z_o = \frac{\omega L}{\omega \sqrt{LC}} = \sqrt{\frac{L}{C}} \tag{8.14}$$

Co-axial cable has a designed impedance of 50 Ω for radio and 75 Ω for TV.

8.2   THE LOSSY TRANSMISSION LINE

If the line is imperfect, there will be some losses due to the resistance of the inner conductor (R) and any leakage from the inner to the outer conductor (a conductance of G).

As shown in Figure 8.3, there are additional components – the series resistance, $R\Delta x$, and the parallel conductance, $G\Delta x$. The solution to the voltage wave equation is

$$V(x,t) = V_o e^{j\omega t} e^{\pm \gamma x} \tag{8.15}$$

where γ is called the propagation coefficient defined as

$$\gamma = \alpha + j\beta \tag{8.16}$$

with α being the attenuation coefficient and β is the phase coefficient as before. The propagation coefficient can be found from

$$\gamma^2 = (R + j\omega L)(G + j\omega C) \tag{8.17}$$

With these definitions we can now write, for a wave travelling in the positive x direction,

$$V(x,t) = V_o e^{-\alpha x} e^{j\omega t} e^{-j\beta x} \tag{8.18}$$

Note that we now have an exponential decay meaning the signal is attenuated as it goes down the line in the positive x direction. We should also note that the reactive components (C and L) dominate the propagation coefficient at high frequencies and so the line becomes almost lossless.

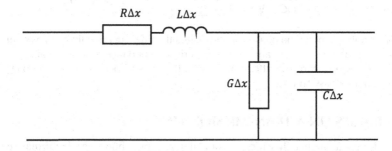

FIGURE 8.3   The lumped parameter model of a lossy transmission line.

The characteristic impedance is given by

$$Z_o = \sqrt{\frac{R + j\omega L}{G + j\omega C}}$$

Example 8.1

A coaxial cable has an inner conductor of diameter 1 mm and an earthed outer conductor of diameter 8 mm. Copper with $\sigma = 54$ MS m^{-1} makes up the inner and outer conductors and the dielectric between the two conductors has $\sigma = 2$ μS m^{-1} with ε_r of 3. Determine the propagation coefficient, the phase velocity and the characteristic impedance all at a frequency of 1 MHz.

Solution

Using (2.37), the lumped parameter capacitance is 80 pF m^{-1}; using (3.51), the total inductance is 466 nH m^{-1}; using (4.18) gives a series resistance of 24 mΩm^{-1}; and use of (4.22) gives a conductance of 6 μS m^{-1}. These values, applied to (8.17), give

$$\gamma^2 = \left(24 \times 10^{-3} + j2.93\right)\left(6 \times 10^{-6} + j0.5 \times 10^{-3}\right)$$

$$= 2.93\underline{/89.5°} \times 5 \times 10^{-4}\underline{/89.3°}$$

$$= 1.5 \times 10^{-3}\underline{/178.8°}$$

$$\text{Therefore,} \quad \gamma = \sqrt{1.5 \times 10^{-3}}\underline{/178.8°/2}$$

$$= 38.7 \times 10^{-3}\underline{/89.4°}$$

$$= 4 \times 10^{-4} + j38.7 \times 10^{-3}$$

i.e., $\alpha = 4 \times 10^{-4}$ m^{-1} and $b = 38.7 \times 10^{-3}$ rad m^{-1}

The phase velocity is $v_p = \dfrac{\omega}{\beta} = \dfrac{2\pi l \times 10^6}{38.7 \times 10^{-3}} = 1.62 \times 10^6$ m s^{-1}

The characteristic impedance is

$$Z_o = \sqrt{\frac{R + j\omega L}{G + j\omega C}} = \sqrt{\frac{2.93\underline{/89.5°}}{5 \times 10 - 4\underline{/89.3°}}} = \sqrt{5860\underline{/0.2°}} = 76.6\underline{/0.1°}\,\Omega$$

Some points of note: the phase velocity is less than the speed of light in a vacuum, the line is almost loss-less and the characteristic impedance is almost real.

We will now turn our attention to what happens to pulses as they propagate down a transmission line.

8.3   PULSES ON A TRANSMISSION LINE

It may be thought that this topic is out of place in a book on electromagnetism. However, the processors used in all computers are running at very high speeds and data rates of 1 Gbit s^{-1} are common. The pulse width of 1 Gbit s^{-1} is 1 ns and that, coupled with the propagation delay along a line, can cause problems as we will see.

Let us consider a transmission line with characteristic impedance Z_o. Let us assume that the load is terminated in a load impedance Z_L. An incident pulse of amplitude V_i gets reflected off the load to give a reflected voltage V_r. (This effect is similar to a radar pulse being reflected off an aircraft. The pulse is travelling in a medium with characteristic impedance 377 Ω. When it meets the metal plane it meets a discontinuity – a short-circuit, i.e., 0 Ω. Some of the pulse is reflected back.) The reflection coefficient, ρ, is given by

$$\rho = \frac{V_r}{V_i} \tag{8.19}$$

The voltage across the load is the sum of the incident and reflected voltages. Thus,

$$V_L = V_i + V_r$$

and so,

$$V_r = V_L - V_i$$
$$= I_L Z_L - V_i \tag{8.20}$$

Now, the current in the load is the sum of the incident and reflected currents but one is entering the load (the incident) and one is leaving (the reflected). So,

$$I_L = I_i - I_r$$

Therefore, (8.20) becomes

$$V_r = (I_i - I_r)Z_L - V_i$$
$$= \left(\frac{V_i}{Z_o} - \frac{V_r}{Z_o}\right)Z_L - V_i$$

After rearranging, this becomes

$$V_r\left(1 + \frac{Z_L}{Z_o}\right) = V_i\left(\frac{Z_L}{Z_o} - 1\right)$$

Therefore, the reflection coefficient, ρ, is

$$\rho = \frac{V_r}{V_i} = \frac{Z_L - Z_o}{Z_L + Z_o} \tag{8.21}$$

A similar derivation yields the transmission coefficient, τ, as

$$\tau = \frac{V_t}{V_i} \frac{2Z_L}{Z_L + Z_o} \tag{8.22}$$

where V_t is the transmitted voltage.

Example 8.2

A 12 V signal is propagating down a 50 Ω transmission line – a 2-mm wide pcb track over a ground plane. The line is terminated in a 1 kΩ load. Determine the voltage at the end of the line.

Solution

The 12 V is travelling in a 50 Ω environment before it sees a discontinuity in the form of the 1 kΩ load. The reflection coefficient is given by

$$\rho = \frac{V_r}{V_i} = \frac{Z_L - Z_o}{Z_L + Z_o}$$

$$= \frac{1000 - 50}{1000 + 50}$$

$$= 0.9 \tag{8.21}$$

The voltage at the load is the sum of the incident signal and the reflected signal. Thus,

$$V_L = 12 + 12 \times 0.9$$

$$= 22.8\,\text{V}$$

This may cause a problem with the load device. The solution is to match to the line by using a load equal to the impedance of the line. Problems 8.3 and 8.4 show the difficulty of designing with high speed pulses.

One very useful way of visualizing what happens to the load voltage is through the use of a lattice diagram.

Example 8.3

A 1 ns, 12 V signal is propagating down a 50 Ω transmission line with a time delay of 10 ns. The line is terminated in a 10 kΩ load and the source impedance is 150 Ω. Show, using a lattice diagram, how the pulse gets reflected off the load and source.

Solution

The 12 V pulse initially sees the 50 Ω transmission line and not the load because the 1 ns pulse takes 10 ns to get to the load. So, the source has a series impedance of 150 Ω, and the line presents a 50 Ω load to the source. Thus, we have a potential divider and the amplitude of the pulse going down the line is

$$V_p = \frac{12}{150 + 50} \times 50 = 3\,\text{V}$$

This pulse travels down the line until it reaches the load of 10 kΩ. Here, it sees an impedance discontinuity and so it experiences a reflection coefficient, ρ_L, of

$$\rho_L = \frac{Z_L - Z_o}{Z_L + Z_o} = \frac{10,000 - 50}{10,000 + 50} = 0.99$$

Thus, the magnitude of the pulse reflected back to the load is 3 × 0.99 = 2.97 V, and the load voltage is 3 + 2.97 = 5.97 V. This may not be a large enough amplitude to be of much use. What has gone wrong is not the fact that the load causes a reflection (it actually helps with amplitude) but the series resistance of the source and the potential divider effect at the source (a 12 V pulse is attenuated to 3 V).

The reflected pulse sees a load of 150 Ω and so the source reflection coefficient, ρ_S, is

$$\rho_S = \frac{Z_S - Z_o}{Z_S + Z_o} = \frac{150 - 50}{150 + 50} = 0.50$$

Thus, the returning pulse has an amplitude of 1.49 V. The pulse bounces back and forward being reflected off the load and source as shown in the lattice diagram (Figure 8.4). The numbers in the brackets are the terminal voltages if the pulse is much longer than the 10 ns line propagation time. We can see that the terminal

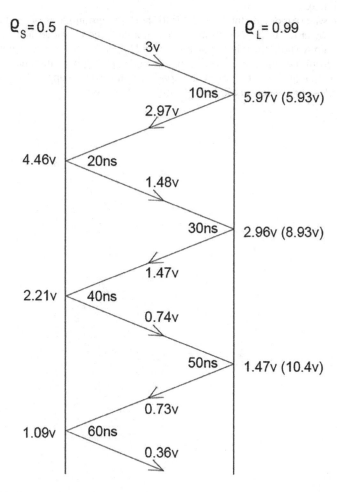

FIGURE 8.4   Lattice diagram for Example 8.3.

voltage is reaching 12 V but only after a considerable amount of time. Thus, it is important to match the load to the line and/or use short lines.

Example 8.4

All the outputs of an octal buffer switch at the same time resulting in a sudden drain in the power supply (Example 8.3). How can this effect be reduced?

Solution

The power rail voltage will drop suddenly when the i.c. requires current to drive the loads. This transient will propagate back to the power supply where a voltage regulator, together with a capacitor, will stop the transient. With a linear power supply, this capacitor will be a large electrolytic sometimes called a 'reservoir' capacitor. Of course, every component that uses the power rail will experience interference.

One way of reducing the transient is to short it to ground as soon as it appears on the power rail using a decoupling capacitor. Its action is rather like what happens when we flush the toilet. Water is sourced locally from the cistern which then fills up from the reservoir. The decoupling capacitor supplies the charge locally before recharging from the reservoir capacitor in the power supply. A capacitor of value 0.1 μF or greater should be sufficient.

9 Maxwell's Equations and Electromagnetic Waves

The work presented in this chapter will enable us to develop the wave equation for electromagnetic waves both in free-space and in a material. We will get involved in some vector analysis but it is introduced in context to help understanding.

9.1 INTEGRAL FORM OF MAXWELL'S EQUATIONS

Gauss' law states that the electrostatic flux flowing through a closed surface is equal to the enclosed charge. We encountered this when we considered electrostatic flux flow in Chapter 2. We can write

$$\varphi = q$$

$$\text{or,} \quad D \times \text{area} = q \tag{9.1}$$

$$\text{and so,} \quad \oint_{\text{surf}} \boldsymbol{D} \cdot \boldsymbol{ds} = q$$

The integral on the left-hand side of (9.1) is the surface integral of the dot product of the flux density, $\boldsymbol{D}$, and the incremental area vector $\boldsymbol{ds}$. The magnitude of vector $\boldsymbol{ds}$ is that of a small area of the Gaussian surface with direction at right-angles to the surface. This is shown in Figure 9.1 which also shows the vector $\boldsymbol{D}$. The dot product is also known as the scalar product and is given by

$$\boldsymbol{a} \cdot \boldsymbol{b} = |a||b|\cos\theta \tag{9.2}$$

where the vertical brackets indicate magnitude and θ is the angle between a and b. The direction of flux through a Gaussian surface (a sphere for point charges or a cylinder for line charges) is the same as for $\boldsymbol{ds}$. So, the angle between them is zero and (9.1) becomes

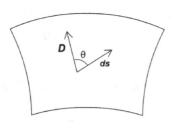

FIGURE 9.1 Relationship between D and ds.

$$\oint_{\text{surf}} D\,ds = q$$

As D is independent of the surface (we are considering a Gaussian surface) the left-hand side is the surface integral of ds which gives the surface area of a Gaussian surface.

The charge can be written as

$$q = \rho \times \text{volume} = \iiint_v \rho\,dv$$

where the triple integral gives the volume and ρ is the charge density in C m^{-3}.

$$\text{Thus} \quad \oint_{\text{surf}} D \cdot ds = \iiint_{\text{vol}} \rho\,dv \qquad (9.3)$$

This is the integral form of Gauss' law of electrostatics and is one of Maxwell's equations.

A similar argument applies to magnetostatics except for there are no isolated monopoles. So the integral form of Gauss' law of magnetostatics is

$$\oint_{\text{surf}} B \cdot ds = 0 \qquad (9.4)$$

Faraday's law of electromagnetic induction relates an induced emf to a time-varying magnetic field. Basically, an emf is induced into a loop of wire if there is a time-varying magnetic field inside the loop. (We come across this in transformers.) Specifically, we have (3.43) for a single turn of wire

$$\text{emf} = -\frac{d\varnothing}{dt} \qquad (9.5)$$

This emf is related to the electric field in the wire loop enclosing the field as

$$\text{emf} = \oint_{\text{loop}} E \cdot dl \qquad (9.6)$$

The magnetic flux is related to the magnetic flux density as

$$\varnothing = \oint_{\text{surf}} B \cdot ds \qquad (9.7)$$

Substituting (9.6) and (9.7) into (9.5) gives

$$\oint_{\text{loop}} E \cdot dl = \frac{d}{dt} \oint_{\text{surf}} B \cdot ds \qquad (9.8)$$

This is the integral form of Faraday's law of electromagnetic induction.

Ampere's law is the last equation we will consider (3.21). This equation links the magnetic field to the current producing it. Thus,

$$\oint_{\text{loop}} H \cdot dl = I + \frac{d\varphi}{dt} \qquad (9.9)$$

In (9.9), we have the familiar electroconduction current and the displacement current that was introduced when we considered capacitors. Current is related to current density and electric flux is related to electric flux density. So, (9.9) becomes

$$\oint_{\text{loop}} H \cdot dl = \oint_{\text{surf}} J \cdot ds + \frac{d}{dt} \oint_{\text{surf}} D \cdot ds \qquad (9.10)$$

This is the integral form of Ampere's law and is the final Maxwell's equations. These four equations need to be manipulated and this is where it can get complicated. Rather than embark on a detailed mathematical analysis, we will discuss the final results in detail.

9.2 DIFFERENTIAL FORM OF MAXWELL'S EQUATIONS

Thus far we have developed four integral form equations. What we would like to do is develop differential equations similar to those we developed for voltage and current on a transmission line. We would then be able to relate the electric and magnetic fields in a similar way to the relationship between voltage and current.

At the heart of the differential, or point, form of Maxwell's equations lies the del operator, ∇. This is a spatial differentiation operator. Let us initially consider Gauss' Law (9.3):

$$\oint_{\text{surf}} D \cdot ds = \iiint_{\text{vol}} \rho \, dv$$

In point form, this is

$$\nabla \cdot D = \rho \qquad (9.11)$$

$\nabla \cdot \boldsymbol{D}$ is the scalar or dot product of ∇ and $\boldsymbol{D}$, and $\nabla \cdot \boldsymbol{D}$ is known as the divergence given by

$$\nabla \cdot \boldsymbol{D} = \frac{\partial D_x}{\partial x} + \frac{\partial D_y}{\partial y} + \frac{\partial D_z}{\partial z} \tag{9.12}$$

As can be seen, the divergence is the rate of change of a vector (in this case the electric flux density, $\boldsymbol{D}$) in three dimensions. It is a scalar quantity, i.e., it only has magnitude. If flux is flowing out of a surface, the surface encloses a positive charge and the divergence is positive. If the charge is negative, the divergence is negative. If there is not net flow of flux, then the divergence is zero.

The corresponding equation for magnetism, assuming there are no magnetic monopoles, is

$$\oint_{\text{surf}} \boldsymbol{B} \cdot \mathbf{ds} = 0$$

which gives

$$\nabla \cdot \boldsymbol{B} = 0 \tag{9.13}$$

As regards the other two equations, these are a little more complicated but very relevant. We have seen the integral form of Faraday's law of electromagnetic induction can be written as

$$\oint_{\text{loop}} \boldsymbol{E} \cdot \mathbf{d}\boldsymbol{l} = -\frac{\mathrm{d}}{\mathrm{d}t} \oint_{\text{surf}} \boldsymbol{B} \cdot \mathbf{ds} \tag{9.14}$$

To generate the differential form of this equation, we must use Stokes' theorem:

$$\oint_{\text{loop}} \boldsymbol{E} \cdot \mathbf{d}\boldsymbol{l} = -\oint_{\text{surf}} \nabla \times \boldsymbol{E} \cdot \mathbf{ds} \tag{9.15}$$

where $\nabla \times \boldsymbol{E}$ (the cross product of ∇ and $\boldsymbol{E}$) is known as the curl of $\boldsymbol{E}$. (It is also known as the circulation of $\boldsymbol{E}$.) Let us take a moment to discuss the curl of the E field.

As shown in (9.14), the E field is integrated around a loop (a loop of wire) and is equal to the rate of change of magnetic field through the surface. The geometry of the situation is shown in Figure 9.2. Stokes' theorem allows us to change the loop integral into a surface integral of the curl of $\boldsymbol{E}$. Thus, (9.14) becomes

$$\oint_{\text{loop}} \boldsymbol{E} \cdot \mathbf{d}\boldsymbol{l} = -\oint_{\text{surf}} \nabla \times \boldsymbol{E} \cdot \mathbf{ds} = \frac{\mathrm{d}}{\mathrm{d}t} \oint_{\text{surf}} \boldsymbol{B} \cdot \mathbf{ds}$$

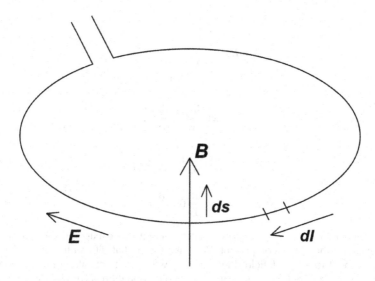

FIGURE 9.2   Showing the production of an electric field from a time-varying magnetic field.

which, after cancelling the surface integrals, gives

$$\nabla \times E = -\frac{d}{dt} B \tag{9.16}$$

The curl of E in Cartesian co-ordinates is

$$\nabla \times E = \begin{vmatrix} x & y & z \\ \dfrac{\partial}{\partial x} & \dfrac{\partial}{\partial y} & \dfrac{\partial}{\partial z} \\ E_x & E_y & E_z \end{vmatrix} \tag{9.17}$$

where x, y and z are unit vectors, the differentials are the partial differentials (only differentiate with respect to one variable, i.e., x, y or z) and E_x, E_y and E_z are the x, y and z components of the E field. An example will be useful.

Example 9.1

An electric field is travelling in the z-direction in free-space with only one x-axis component. Determine the direction of the magnetic field.

Solution

We only have one E field component and it is acting along the x-axis and so (8.35) becomes

$$\nabla \times \boldsymbol{E} = \begin{vmatrix} \boldsymbol{x} & \boldsymbol{y} & \boldsymbol{z} \\ \dfrac{\partial}{\partial x} & \dfrac{\partial}{\partial y} & \dfrac{\partial}{\partial z} \\ E_x & 0 & 0 \end{vmatrix}$$

$$= \boldsymbol{x}\left(\frac{\partial}{\partial y}0 - \frac{\partial}{\partial z}0\right)$$

$$-\boldsymbol{y}\left(\frac{\partial}{\partial x}0 - \frac{\partial}{\partial z}E_x\right)$$

$$+\boldsymbol{z}\left(\frac{\partial}{\partial x}0 - \frac{\partial}{\partial y}E_x\right)$$

We appear to have two components here, along the y-axis and the z-axis. Let us examine the z-axis component. We have the partial differential of the E field with respect to y. The E field does not vary with y – it acts along the x-axis and travels in the z-direction. Hence, this partial differentiation is zero leaving only the differentiation with respect to z – the direction of propagation. So, the H field acts along the y-axis. Thus, we have a three axis relationship – the E field acts along the x-axis, the H field acts along the y-axis and they both travel along the z-axis.

Let us now turn to Ampere's law (9.10)

$$\oint_{loop} \boldsymbol{H} \cdot d\boldsymbol{l} = \oint_{surf} \boldsymbol{J} \cdot d\boldsymbol{s} + \frac{d}{dt}\oint_{surf} \boldsymbol{D} \cdot d\boldsymbol{s}$$

Application of Stokes' theorem gives

$$\oint_{loop} \boldsymbol{H} \cdot d\boldsymbol{l} = \oint_{surf} \nabla \times \boldsymbol{H} d\boldsymbol{s} = \oint_{surf} \boldsymbol{J} \cdot d\boldsymbol{s} + \frac{d}{dt}\oint_{surf} \boldsymbol{D} \cdot d\boldsymbol{s}$$

and so,

$$\nabla \times \boldsymbol{H} = \boldsymbol{J} + \frac{d}{dt}\boldsymbol{D} \tag{9.18}$$

Table 9.1 is a summary of Maxwell's equations both integral and point form.

TABLE 9.1
Summary of Maxwell's Equations in Integral and Point Form

	Integral Form	Differential Form
Gauss' law of electrostatics	$\oint_{surf} \boldsymbol{D} \cdot d\boldsymbol{s} = \iiint_{vol} \rho \, dv$	$\nabla \cdot \boldsymbol{D} = \rho$
Gauss' law of magnetostatics	$\oint_{surf} \boldsymbol{B} \cdot d\boldsymbol{s} = 0$	$\nabla \cdot \boldsymbol{B} = 0$
Faraday's law of induction	$\oint_{loop} \boldsymbol{E} \cdot d\boldsymbol{l} = -\frac{d}{dt}\oint_{surf} \boldsymbol{B} \cdot d\boldsymbol{s}$	$\nabla \times \boldsymbol{E} = -\frac{d}{dt}\boldsymbol{B}$
Ampere's circuital law	$\oint_{loop} \boldsymbol{H} \cdot d\boldsymbol{l} = \oint_{surf} \boldsymbol{J} \cdot d\boldsymbol{s} + \frac{d}{dt}\oint_{surf} \boldsymbol{D} \cdot d\boldsymbol{s}$	$\nabla \times \boldsymbol{H} = \boldsymbol{J} + \frac{d}{dt}\boldsymbol{D}$

9.3 ELECTROMAGNETIC WAVE PROPAGATION IN FREE-SPACE

Maxwell's third equation links the electric field to the magnetic field, whereas the fourth equation links the magnetic field to the electric field. Thus, one field gives rise to another – they are interdependent. In this section, we will encounter some rather complex mathematics which can be neglected on a first reading. It is the final result that is important.

Consider Maxwell's third equation, (9.16) and take the curl of the left-hand side. Vector algebra expands the curl of a curl as

$$\nabla \times \nabla \times E = \nabla(\nabla \cdot E) - \nabla^2 E \qquad (9.19)$$

If we are considering fields in free-space, there are no charges present and so Maxwell's first law yields $\nabla \cdot D = 0$ and so $\nabla \cdot E = 0$. Hence, (9.19) becomes

$$\nabla \times \nabla \times E = -\nabla^2 E \qquad (9.20)$$

Application of Maxwell's third law (relating the curl of E to the time derivative of B) yields

$$\nabla \times \left(-\frac{dB}{dt} \right) = -\nabla^2 E$$

and so,

$$-\frac{d}{dt}(\nabla \times B) = -\nabla^2 E$$

which becomes

$$\mu_o \frac{d}{dt}(\nabla \times H) = \nabla^2 E$$

Maxwell's fourth equation gives the curl of H in terms of the time derivative of the D field. Thus,

$$\mu_o \frac{d}{dt}\left(\frac{dD}{dt} \right) = \nabla^2 E$$

and so,

$$\varepsilon_0 \mu_o \frac{d}{dt}\left(\frac{dE}{dt} \right) = \nabla^2 E$$

or

$$\nabla^2 E = \varepsilon_0 \mu_o \frac{d^2 E}{dt^2} \tag{9.21}$$

The $\nabla^2 E$ term is known as the Laplacian given by

$$\nabla^2 E = \frac{\partial^2 E}{\partial x^2} + \frac{\partial^2 E}{\partial y^2} + \frac{\partial^2 E}{\partial z^2} \tag{9.22}$$

Let us assume the E field acts along the y-axis and travels along the z-axis. Further, let us assume the E field varies sinusoidally with time. The wave equation of (9.21) becomes

$$\frac{\partial^2 E}{\partial z^2} = \varepsilon_0 \mu_o \frac{d^2 E}{dt^2} \tag{9.23}$$

Comparison with the voltage wave equation (8.7) reveals a great deal of similarity and so a possible solution is

$$E = E_y \cos(\omega t - \beta z) y \tag{9.24}$$

A similar derivation for the magnetic field gives

$$H = H_x \cos(\omega t - \beta z) x \tag{9.25}$$

So, we have two fields at right angles to each other in a plane that is at right angles to the direction of propagation (Figure 9.3). We can treat these travelling waves in the same way as the travelling voltage and current waves in the last chapter. Substitution of (9.24) into (9.23) yields (this is left as an exercise for the reader)

$$\beta = \omega \sqrt{\mu_o \varepsilon_o} \tag{9.26}$$

The phase velocity is obtained by plotting (9.24) in a similar way as the voltage travelling wave

$$v_p = \frac{1}{\sqrt{\mu_o \varepsilon_o}} = 3 \times 10^8 \text{ m s}^{-1} \tag{9.27}$$

This is the speed of light and is the reason why light was considered to be an electromagnetic wave. Substitution of the E and H solutions into Maxwell's third equation gives the impedance of free-space as

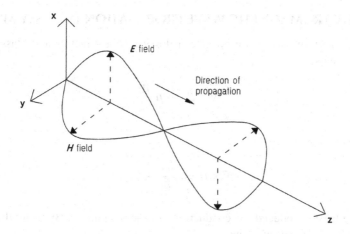

FIGURE 9.3   Variation of E and H for a TEM wave propagating in the z-direction.

$$Z = \frac{E}{H} = \sqrt{\frac{\mu_o}{\varepsilon_o}} = 377 \ \Omega$$

Note that if the medium is not free-space but a dielectric, the relative permittivity must be included. And so,

$$v_p = \frac{1}{\sqrt{\mu_o \varepsilon_o \varepsilon_r}} = \frac{3 \times 10^8}{n} \text{m s}^{-1} \qquad (9.28)$$

where n is the refractive index given by

$$n = \sqrt{\varepsilon_r} \qquad (9.29)$$

An electromagnetic wave conveys power and this is represented by the Poynting vector, S,

$$S = E \times H^* \qquad (9.30)$$

where H^* is the complex conjugate of the magnetic field. It should be noted that these are peak values of E and H. If we work with r.m.s. values, we will get the average power

$$S_{av} = \frac{1}{2} E \times H^* \qquad (9.31)$$

9.4   ELECTROMAGNETIC WAVE PROPAGATION IN LOSSY MEDIA

Electromagnetic waves also propagate in lossy media – metals and glass. In this instance, we use

$$\nabla \times E = -\mu \frac{\partial H}{\partial t} \tag{9.32}$$

and

$$\nabla \times H = \varepsilon \frac{\partial E}{\partial t} + \sigma E \tag{9.33}$$

where we have introduced the conductivity of the medium, σ. By using the phasor forms of E and H, we can write

$$E = E_x \exp(jwt)x \qquad \text{and} \qquad H = H_y \exp(jwt)y$$

We can now write (9.32) and (9.33) as

$$\frac{\partial E}{\partial z} = -j\omega\mu H \tag{9.34}$$

$$\text{and} \quad -\frac{\partial H}{\partial z} = j\omega\varepsilon E + \sigma E \tag{9.35}$$

These two equations can be manipulated in the usual way by differentiating (9.34) with respect to z and substituting from (9.35). So,

$$\frac{\partial^2 E}{\partial z^2} = -\omega^2 \mu\varepsilon E + j\omega\mu E \tag{9.36}$$

Similarly,

$$\frac{\partial^2 H}{\partial z^2} = -\omega^2 \mu\varepsilon H + j\omega\mu H \tag{9.37}$$

If we now let $\gamma^2 = -\omega^2 \mu\varepsilon + j\omega\mu$, possible solutions to these equations are

$$E = E_{xo} \exp(j\omega t)\exp(-\gamma z)x \tag{9.38}$$

and

$$H = H_{yo} \exp(j\omega t)\exp(-\gamma z)y \tag{9.39}$$

where the subscript o denotes the values of E and H at the origin of a right handed Cartesian coordinate set, and γ is known as the propagation coefficient given by $\gamma = \alpha + j\beta$, where α and β are the attenuation and phase coefficients, respectively, we get

$$E = E_{xo}e(-z)\cos(\omega t - \beta z)x \qquad (9.40)$$

and

$$H = H_{yo}e(-z)\cos(\omega t - \beta z)y \qquad (9.41)$$

Here, we have the familiar travelling wave propagating in the positive z-direction, undergoing attenuation as $\exp(-\alpha z)$.

Example 9.2

Determine the attenuation and phase coefficients at a frequency of 1 MHz for glass ($\sigma = 1 \times 10^{-13}$ S m^{-1}, $\varepsilon_r = 4$, $\mu_r = 1$) and copper ($\sigma = 59.6$ MS m^{-1}, $\varepsilon_r = 1$, $\mu_r = 1$).

Solution

The propagation coefficient, $\gamma = \alpha + j\beta$, can be found from $\gamma^2 = -\omega^2 \mu\varepsilon + j\omega\sigma\mu$. For the glass, the imaginary term is of the order of 10^{-13}, while the real term is of the order of 10^{-6}. Thus,

$$\gamma^2 = -\omega^2 \mu\varepsilon$$

and therefore

$$\gamma = j\omega\sqrt{\mu\varepsilon}$$

which means $\alpha = 0$ and $\beta = \omega\sqrt{\mu\varepsilon} = 0.041$. This implies zero attenuation or, at the very least, very low attenuation. This is put to good use in optical fibre communication links.

For the copper, the imaginary term is dominant because σ is so large, and so

$$\gamma^2 = j\omega\mu\sigma$$

Now, j is $1/90°$ and so $\sqrt{j} = 1/45°$. Thus,

$$\gamma = \sqrt{\frac{\omega\mu}{2}} + j\sqrt{\frac{\omega\mu\sigma}{2}}$$

Therefore, $\alpha = \beta = 15{,}340$. Thus, copper has a very large loss. This is put to good use in electromagnetic shielding.

9.5 SKIN EFFECT

This is a fascinating effect that has a bearing on conductor design at all frequencies. The easiest way to study the phenomenon is to consider what happens to an electric field as it propagates through a metal.

The example in the last section examined how electromagnetic fields are absorbed by metals. It was found that

$$\alpha = \sqrt{\frac{\omega\mu}{2}} \tag{9.42}$$

and that the electromagnetic field reduces as $e^{-\alpha x}$ where x is the distance travelled into the metal. The distance at which the attenuation is $1/e$ of the surface value is known as the skin depth, δ. So,

$$\delta = \frac{1}{\alpha}$$

$$= \sqrt{\frac{2}{\omega\mu}} \tag{9.43}$$

An analogy can be drawn with the exponential decay associated with time constants – after five time constants the transient has passed. Here, we can say that after five skin depths, the field is approximately zero. As $D = \varepsilon E$ and $J = \sigma E$ the distribution of current throughout the cross-section of a conductor is non-uniform – it tends to stay near the surface, the skin, and decays to zero after 5δ.

Example 9.3

Determine the skin depth for copper, $\sigma = 54$ MS m^{-1}, at a frequency of 50 Hz and hence estimate the maximum diameter of a conductor.

Solution

Application of (9.41) yields $\delta = 9.2$ mm, and so the conductor diameter can be $2 \times 5 \times \delta = 9.2$ cm.

The fact that current tends to flow in the skin of a conductor means that the a.c. resistance is greater than the d.c. one. It is not possible to reduce the a.c. resistance by simply increasing the conductor size. For frequencies above 1 MHz, the wire can be silver plated because the skin depth is small in silver. For microwave frequencies, the inner portion can be completely removed and we are left with a hollow conductor called a waveguide. In this case, it is common to consider the field in the waveguide rather than the current in the wall of the guide.

9.6 COMMENT

Maxwell's equations have resulted in several conclusions:

- A time-varying magnetic field gives rise to a time-varying electric field and vice versa
- These time-varying fields are called electromagnetic fields

- In free-space, the fields propagate with the speed of light
- Light is an electromagnetic wave

At the time Maxwell published his theory (1864), physicists believed in the aether – a fluid that surrounded all things. We are used to sound waves relying on air to propagate and waves rely on water. So, why not electromagnetic waves using the aether?

In 1887, two physicists – Michelson and Morley – performed an experiment to determine the drift of the Earth through the moving aether. They reasoned that the speed of light in one direction would be faster if travelling with the aether, in the same way that the speed of sound increases if the source is moving in the direction of the sound wave. Then, by rotating the experiment, the speed of light would be lower in the same way that the speed of the sound wave decreases as the source moves away. Their experiment gave a null result – the speed of light remained constant regardless of the orientation of the equipment. There then followed many theories that attempted to justify the null experiment and keep the aether.

In 1905, almost 20 years later, it was Einstein who postulated that the speed of light is a constant regardless of the frame of reference. This was the Special Theory of Relativity and it contains some rather challenging ideas. Consider a light pulse emitted at point A in a stationary spacecraft. It is reflected off the ceiling at point B and so arrives back at point A sometime later (Figure 9.4). The speed of propagation is simply the distance travelled (floor to ceiling twice) divided by the time of flight. To put some numbers to the problem, let us take a ceiling height of 6 m. Thus, the round trip takes

$$\frac{2 \times 6}{c}$$

$$= \frac{12}{3 \times 10^8}$$

$$= 40 \, \text{ns}$$

So, relative to the spacecraft the light travels at the speed of light and it is in flight for 40 ns. Now consider what happens if the spacecraft is moving relative to an observer on the ground. In the 40 ns taken for the light pulse to reach the ceiling and return, the spacecraft will have travelled to the left by a certain amount. If we take the velocity of the spacecraft to be half the speed of light, this certain amount could be considerable. So, the light pulse will not hit point B because it has moved to the left while the pulse is in flight. Instead, it will hit the ceiling at C before returning to a different point, point D. It should be obvious that the light pulse has travelled further taking path ACD, instead of the direct route path AB. The observer in the spacecraft sees the light pulse bounce off point B, whereas the observer on the ground sees a longer path for exactly the same event. Two of the postulates that Einstein used were that the laws of physics hold whichever frame is being considered, and that the speed of light is a constant. The path ACD is longer and so the time taken, t, must be longer to keep the speed of light constant.

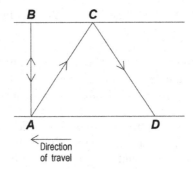

FIGURE 9.4 Propagation of a light pulse for an observer aboard a spacecraft and an observer on the ground.

The spacecraft is travelling at $c/2$ and so the distance travelled in time t is

$$2x = 1.5 \times 10^8 t$$

Therefore, $x = 0.75 \times 10^8 t$

The pulse follows the path ACD which is $2y$ long. The speed of travel is that of light and so,

$$2y = 3 \times 10^8 t$$

Thus, $y = 1.5 \times 10^8 t$

We have a right-angled triangle and so,

$$y^2 = x^2 + 6^2$$

Substituting for x and y gives

$$\left(1.5 \times 10^8 t\right)^2 - \left(0.75 \times 10^8 t\right)^2 = 36$$

And so $t = 46$ ns. We have not been able to measure the speed of light directly on a spacecraft travelling at half the speed of light. Also, the effect of gravity has been ignored. (This is the topic of the General Theory of Relativity.) However, there has been experimental verification that time slows down at high speeds. Atomic clocks were sent around the Earth in eastward and westward directions. Differences were found in the times of the clocks when they were compared to an atomic clock on the ground with which they were originally synchronized. The Global Positioning System (GPS) or SatNav relies on the theory of relativity to get location information.

So, there we are. Time passes slower as speed increases; there is no aether; there are no magnetic monopoles. A Nobel Prize surely awaits the discoverer of the latter two. Happy hunting!

10 The Planar Optical Waveguide

In this chapter, we will consider Maxwell's equations as applied to a planar optical waveguide. We will start by considering what happens when light is reflected off a boundary.

10.1 REFLECTION AT BOUNDARIES

We have already seen in the last chapter that signals can be reflected at changes in impedance. The same is true of electromagnetic waves – we get reflection and transmission at the boundary of two media. A familiar example of this is a window. When we look through a pane of glass, we see what is behind the glass (transmission) and what is in front (reflection). What is happening is that the light sees a change in impedance giving reflection and transmission. To study this in greater detail, let us consider an electromagnetic wave incident on a boundary between dissimilar materials as shown in Figure 10.1.

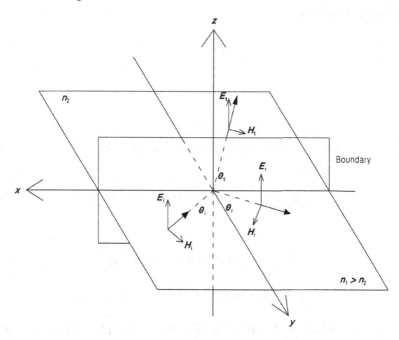

FIGURE 10.1 Reflection and refraction of a TEM wave at the boundary between two dielectric materials. (Reprinted by permission from *Optical Communications*, Springer by Martin J. N. Sibley © 2019.)

Here, the incident wave has the subscript i, the reflected the subscript r and the transmitted subscript t. Consider the E field. The incident field acts at an angle θ_i, and we can resolve the incident field into x and y components. Thus,

$$E_i = E_i \exp\left(j\beta_1 \left[x\sin\theta_i + y\cos\theta_i \right]\right) z \qquad (10.1)$$

The reflected and transmitted fields can be written as

$$E_r = E_r \exp\left(j\beta_1 \left[x\sin\theta_r + y\cos\theta_r \right]\right) z \qquad (10.2)$$

$$E_t = E_t \exp\left(j\beta_2 \left[x\sin\theta_t + y\cos\theta_t \right]\right) z \qquad (10.3)$$

where x and y are the distances travelled along the respective axes. We are now in a position to apply the boundary conditions and so generate relationships between the incident and reflected and transmitted fields.

As found in Section 6.3, the tangential component of the E field is constant across the boundary. Thus, the sum of the E fields in medium 1 must equal the sum of the E fields in medium 2. So,

$$E_i + E_r = E_t \qquad (10.4)$$

Dividing throughout by E_i gives

$$1 + \frac{E_r}{E_i} = \frac{E_t}{E_i}$$

and so,

$$1 + \rho = \tau \qquad (10.5)$$

where ρ is the reflection coefficient and τ is the transmission coefficient. So, E_r and E_t become

$$E_r = \rho E_i \exp\left(j\beta_1 \left[x\sin\theta_r + y\cos\theta_r \right]\right) z \qquad (10.6)$$

and

$$E_t = \tau E_i \exp\left(j\beta_2 \left[x\sin\theta_t + y\cos\theta_t \right]\right) z \qquad (10.7)$$

At the interface, $y = 0$, substitution of (10.6) and (10.7) into (10.4) gives

$$E_i \exp\left(j\beta_1 x\sin\theta_i\right) + \rho E_i \exp\left(j\beta_1 x\sin\theta_r\right) = \tau E_i \exp\left(j\beta_2 x\sin\theta_t\right) \qquad (10.8)$$

Comparison with Equation (10.5) indicates that the exponential terms in (10.8) must be equal to each other so that they can cancel out and satisfy (10.5), that is

$$\beta_1 \sin\theta_i = \beta_1 \sin\theta_r = \beta_2 \sin\theta_t$$

The first of these equalities gives Snell's law of reflection,

$$\theta_i = \theta_r \tag{10.9}$$

The second equality gives Snell's law of refraction,

$$\sin\theta_t = \frac{\beta_1}{\beta_2}\sin\theta_i = \frac{n_1}{n_2}\sin\theta_i \tag{10.10}$$

We need to generate another equation to find expressions for ρ and τ. To do this, we consider the continuity of the tangential magnetic field. The H fields act at right angles to their respective directions of propagation, and so we can write

$$H_i = \left(-\cos\theta_i x + \sin\theta_i y\right)\frac{E_i}{Z_1} \tag{10.11}$$

$$H_r = \left(\cos\theta_r x + \sin\theta_r y\right)\frac{E_r}{Z_1} \tag{10.12}$$

$$H_t = \left(-\cos\theta_t x + \sin\theta_t y\right)\frac{E_t}{Z_2} \tag{10.13}$$

The continuity of the tangential H field at the boundary gives, at $y = 0$

$$\frac{E_i}{Z_1}\cos\theta_i \exp(j\beta_{1x}x) - \rho\frac{E_i}{Z_1}\cos\theta_r \exp(j\beta_{1x}x) = \tau\frac{E_i}{Z_2}\cos\theta_t \exp(j\beta_{2x}x) \tag{10.14}$$

In this equation, the parameters β_{1x} and β_{2x} are the phase constants for materials 1 and 2 resolved onto the x-axis

$$\beta_{1x} = \beta_1 \sin\theta_i = \beta_1 \sin\theta_r \tag{10.15}$$

and

$$\beta_{2x} = \beta_2 \sin\theta_t \tag{10.16}$$

When we considered the E field, the exponential terms in (10.14) were all equal. Therefore,

$$\frac{\cos\theta_i}{Z_1}(1 - \rho) = \frac{\tau\cos\theta_t}{Z_2} \tag{10.17}$$

Since $1 + \rho = \tau$, we can eliminate τ from (10.17) to give

$$\rho = \frac{Z_2\cos\theta_i - Z_1\cos\theta_t}{Z_2\cos\theta_i + Z_1\cos\theta_t} = \frac{n_1\cos\theta_i - n_2\cos\theta_t}{n_1\cos\theta_i + n_2\cos\theta_t} \qquad (10.18)$$

θ_i can be eliminated from (10.18) by using Snell's law and so,

$$\rho = \frac{\cos\theta_i - \sqrt{\left(\dfrac{n_2}{n_1}\right)^2 - \sin^2\theta_i}}{\cos\theta_i + \sqrt{\left(\dfrac{n_2}{n_1}\right)^2 - \sin^2\theta_i}} \qquad (10.19)$$

Total internal reflection occurs when $\rho = 1$, and this occurs when the root term is zero, i.e., $\sin^2\theta_i = (n_2/n_1)^2$. This angle is called the critical angle, θ_c, given by

$$\sin^2\theta_c = \left[\frac{n_2}{n_1}\right]^2 \text{ or } \sin\theta_c = \frac{n_2}{n_1} \qquad (10.20)$$

Use of this angle in Snell's law gives an angle of refraction of 90°. This means that there is a transmitted ray travelling along the boundary. The critical angle is the minimum angle for which total internal reflection occurs. If θ_i is greater than θ_c, ρ will be complex but $|\rho|$ will be unity, total internal reflection still takes place and there is a transmitted wave. We have already seen that the transmitted E field is given by

$$E_t = \tau E_i \exp\left(j\beta_{2x}x + j\beta_{2y}y\right)z \qquad (10.21)$$

To evaluate E_t, we need to evaluate $\sin\theta_t$ and $\cos\theta_t$ and hence β_{2x} and β_{2y}. If $\theta_i > \theta_c$, then $\sin\theta_i > n_2/n_1$ and substitution into Snell's law gives $\sin\theta_t > 1$. If we let $\sin\theta_t > 1$, $\cos\theta_t$ will be imaginary:

$$\cos\theta_t = \sqrt{1 - \sin^2\theta_t} = j\sqrt{(n_1/n_2)^2 \sin^2\theta_i - 1} = jA \qquad (10.22)$$

Thus, the transmitted wave (10.21) can be written as

$$E_t = \tau E_i \exp\left(j\beta_{2x}x + j\beta_{2y}y\right)z$$

$$= \tau E_i \exp\left(j\beta_{2x}x + j^2 A\beta_2 y\right)z$$

$$= \tau E_i \exp\left(-A\beta_2 y\right)\exp\left(j\beta_{2x}x\right)z \qquad (10.23)$$

So, an E field propagates along the x-axis but is attenuated exponentially at right angles along the y-axis. To find the transmitted power, we also need an expression for the transmitted H field. This field is given by Equation (10.13)

$$H_t = \left(-\cos\theta_t \mathbf{x} + \sin\theta_t \mathbf{y}\right)\frac{E_t}{Z_2}$$

where $E_t = \tau E_o \exp\left(-A\beta_2 y\right)\exp\left(j\beta_{2x}x\right)$. Substitution for $\cos\theta_t$ yields

$$H_t = \left(-jA\mathbf{x} + \sin\theta_t \mathbf{y}\right)\frac{E_t}{Z_2} \tag{10.24}$$

Thus, there is a component along the negative x-axis and a component along the y-axis. This means that there will be two components to the power. The average power is given by

$$S_{av} = \frac{1}{2}E \times H^*$$

where $H^* = H \exp\left(-j[\omega t - \Phi]\right)$ and Φ is the temporal phase shift between the E and H field components. This phase shift is $\pi/2$ for the x-axis component of H_t and so

$$S_{av} = -\frac{1}{2}\times\frac{E_t^2}{Z_2}\sin\theta_t \mathbf{x} - \frac{1}{2}\times\frac{AE_t^2}{Z_2}\exp\left(j\pi/2\right)\mathbf{y}$$

$$= -\frac{1}{2}\times\frac{E_t^2}{Z_2}\sin\theta_t \mathbf{x} - j\frac{1}{2}\times\frac{AE_t^2}{Z_2}\mathbf{y} \tag{10.25}$$

As Equation (10.25) shows, there are two components to the transmitted power: an x-axis component acting along the boundary and an imaginary component at right angles to the boundary. The interpretation of this imaginary component is that for the first quarter cycle the E field is positive and the H field is negative (90° phase shift) so the power flow is negative. In the next quarter cycle, the E field is still positive but the H field is also positive giving positive power flow. This power flows into and away from the boundary results in no net flow of power along the y-axis.

The real part of the Poynting vector is

$$S_{av} = -\frac{1}{2}\times\frac{E_t^2}{Z_2}\sin\theta_t \mathbf{x} \tag{10.26}$$

which shows that power flows *along* the boundary although total internal reflection takes place. This is known as the evanescent wave and, as shown in the following example, it is very tightly bound to the boundary through the two media.

Example 10.1

Light, with a free-space wavelength of 633 nm, is propagating in a dielectric of refractive index 1.4. The wave hits a boundary with a second dielectric of refractive index 1.2 at an angle of 80° to a normal drawn perpendicular to the boundary. Determine the average transmitted power and calculate the attenuation of the evanescent wave at a distance of one wavelength from the boundary.

Solution

Let us initially calculate the reflection coefficient from which we can find the transmission coefficient. Now, ρ is given by (10.19)

$$\rho = \frac{\cos\theta_i - \sqrt{\left(\frac{n_2}{n_1}\right)^2 - \sin^2\theta_i}}{\cos\theta_i + \sqrt{\left(\frac{n_2}{n_1}\right)^2 - \sin^2\theta_i}}$$

$$= \frac{0174 - \sqrt{0.735 - 0.970}}{0.174 + \sqrt{0.735 - 0.970}}$$

$$= \frac{0.174 - j0.485}{0.174 + j0.485}$$

$$= 1/-70.3°$$

Since $\tau = 1 + \rho$

$$t_e = \frac{0.174 + j0.485}{0.174 + j0.485} + \frac{0.174 - j0.485}{0.174 + j0.485}$$

$$= 2 \times 0.174/(0.174 + j0.485) = 0.675/-35.2°$$

Thus, E_t (10.23) will be

$$E_t = \tau E_i \exp\left(-A\beta_2 y\right)\exp\left(j\beta_{2x}x\right)z$$

A is given by Equation (10.22)

$$A = \sqrt{\left(n_1/n_2\right)^2 \sin^2\theta_i - 1} = 0.566$$

and β_2 is given by Equation (10.16)

$$\beta_2 = \frac{\omega}{cn_2} = \frac{2\pi}{\lambda n_2} = 8.27 \times 10^6 \text{rad m}^{-1}$$

Thus,

$$E_t = 0.675 E_i \exp\left(-4.68 \times 10^6 y\right)\exp\left(j8.27 \times 10^6 x\right)z$$

Hence, the average transmitted power is (E^2/Z with $Z = 377/n_2$)

$$S_{av} = 5.0\times10^{-4} E_i^2\exp\left(-4.68\times10^6 y\right)$$

We should note that, as the magnitude of ρ is unity, the reflected power is the same as the incident power. (The angle associated with ρ is simply the spatial phase shift experienced at the boundary.)

If we consider the average power at y equals one wavelength in the second dielectric, we find

$$S_{av} = 5.00\times10^{-4} E_i^2\exp(-2.47) \quad \text{for } y = 528\,\text{nm}$$

$$= 4.23\times10^{-5} E_i^2$$

This attenuation is 43.74 dB.

10.2 PROPAGATION IN PLANAR OPTICAL WAVEGUIDES

Planar optical waveguides are widely used in optical communications as modulators and integrated optical devices. They consist of a slab of high-refractive-index material surrounded by material of lower refractive index (see Figure 10.2). There are two ways of studying propagation in planar waveguides – a ray path analysis and application of Maxwell's equations together with boundary relationships. This section will tackle both analyses.

10.2.1 RAY PATH ANALYSIS

Figure 10.3 shows two light rays propagating in a planar optical waveguide. The electric fields are drawn at right-angles to the direction of propagation. In order to propagate successfully, these E fields have to be in phase so that they add constructively. If they are not in phase, they will add destructively and the ray will be lost.

In order to maintain constructive interference, the phase change from A to B must be an integral number of cycles. The ray crosses the waveguide twice and is reflected off the boundary twice in travelling from A to B. The reflection coefficient, ρ, is (10.19)

$$r_e = \frac{\cos\theta_i - \sqrt{\left(\dfrac{n_2}{n_1}\right)^2 - \sin^2\theta_i}}{\cos\theta_i + \sqrt{\left(\dfrac{n_2}{n_1}\right)^2 - \sin^2\theta_i}}$$

or

$$r_e = 1/-2\phi \quad \text{where} \quad \phi = \tan^{-1}\frac{\sqrt{\left(n_1^2\sin^2\theta_i - n_2^2\right)}}{n_1\cos\theta_i} \tag{10.27}$$

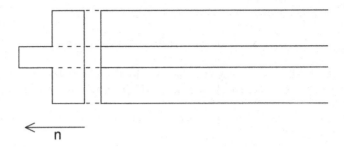

FIGURE 10.2 Refractive index profile of a planar optical waveguide.

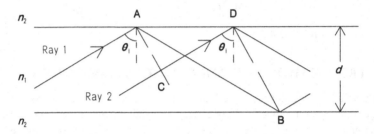

FIGURE 10.3 Illustrative of the requirement for successful propagation of two TEM waves in a planar optical waveguide (CD == a, AB == b). (Reprinted by permission from *Optical Communications*, Springer by Martin J. N. Sibley © 2019.)

Therefore, successful propagation occurs provided

$$2 \times 2d \times \beta_{1y} + 2 \times 2\phi = 2\pi N \tag{10.28}$$

where β_{1y} is the phase coefficient resolved onto the y-axis and N is a positive integer, known as the *mode number*. (It should be noted that two rays make up a single waveguide mode – an upward travelling ray and a downward travelling ray.)

If we substitute for ϕ Equation (10.28) becomes

$$2d \times \beta_{1y} - 2\tan^{-1} \frac{\sqrt{\left(n_1^2 \sin^2\theta_i - n_2^2\right)}}{n_1 \cos\theta_i} = \pi N$$

or

$$\tan\left(\beta_{1y}d - \frac{\pi}{2}N\right) = \frac{\sqrt{\left(n_1^2 \sin^2\theta_i - n_2^2\right)}}{n_1 \cos\theta_i} \tag{10.29}$$

Now, $\beta_{1y} = \beta_1 \cos\theta_i$ and $\beta_1 = k_o n_1$ where k_o is the free-space phase coefficient $\left(k_o = \dfrac{2\pi}{\lambda_o}\right)$ and so we can write Equation (10.29) as

$$\tan\left(\beta_{1y}d - \frac{\pi}{2}N\right) = \frac{2\pi\sqrt{\left(n_1^2\sin^2\theta_i - n_2^2\right)}}{\beta_{1y}\lambda_o} \tag{10.30}$$

As we have already seen, the evanescent field is exponentially attenuated $\exp(-\alpha_2 y)$ and, from (10.23), we can write the attenuation factor as

$$\alpha_2 = \beta_2 A$$

$$= \beta_2\sqrt{\left(n_1/n_2\right)^2\sin^2\theta_i - 1}$$

or

$$\alpha_2 = \frac{2\pi\sqrt{\left(n_1^2\sin^2\theta_i - n_2^2\right)}}{\lambda_o} \tag{10.31}$$

Therefore, we can write (10.30) as

$$\tan\left(\beta_{1y}d - \frac{\pi}{2}N\right) = \frac{\alpha_2}{\beta_{1y}} \tag{10.32}$$

Both Equations (10.29) and (10.32) are known as *eigenvalue* equations with the solution to (10.32) giving the values of β_{1y}, the *eigenvalues*, for which rays will propagate, while solution of (10.29) yields the permitted values of θ_i. Unfortunately, we can only solve these equations using graphical or numerical methods as the following example shows.

Example 10.2

Light of wavelength 1.3 µm is propagating in a planar waveguide of width 200 µm, depth 10 µm and refractive index 1.46, surrounded by material of refractive index 1.44. Find the permitted angles of incidence.

Solution

We can find the angle of incidence for each propagating mode by substituting these parameters into Equation (10.32) and then solving the equation by graphical means. This is shown in Figure 10.4, which is a plot of the left- and righthand sides of Equation (10.32), for varying angles of incidence, θ_i.

This graph shows that approximate values of θi are 88°, 86°, 84° and 82°, for mode numbers 0–3, respectively. Taking these values as a starting point, we can use numerical iteration to find the values of θi to any degree of accuracy. Thus, the values of θi, to two decimal places, are 87.83°, 85.67°, 83.55° and 81.55°.

So, only those modes that satisfy (10.32) are allowed to propagate in the waveguide. As shown in Example 10.4, the higher order modes have the smallest angles of incidence. The limit is the critical angle – modes with angle less than θ_c will not propagate. We can use θ_c in (10.32) to give

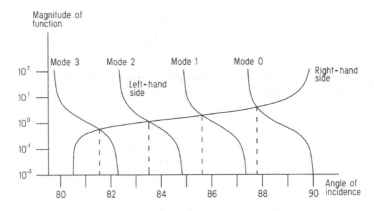

FIGURE 10.4 Eigenvalue graphs for a planar dielectric waveguide. (Reprinted by permission from *Optical Communications*, Springer by Martin J. N. Sibley © 2019.)

$$\left(\frac{2\pi n_1}{\lambda_o} d\cos\theta_c - \frac{\pi}{2} N_{\text{max}} \right) = 0$$

or

$$\frac{2\pi d \sqrt{n_1^2 - n_2^2}}{\lambda_o} = N_{\text{max}} \frac{\pi}{2}$$

where N_{max} is the maximum mode order. We can define a normalized frequency variable, V, as

$$V = \frac{2\pi d \sqrt{n_1^2 - n_2^2}}{\lambda_o} \tag{10.33}$$

then the maximum number of modes will be

$$N_{\text{max}} = \frac{2V}{\pi} = \frac{4d \sqrt{n_1^2 - n_2^2}}{\lambda_o} \tag{10.34}$$

The value N_{max} is unlikely to be an integer and so we must round it up to the nearest whole number. If we take the previous example, then $V = 5.82$ and so the number of propagating modes is 4 which is the same as found in the previous example. We can find the condition for single-mode operation from V. If $N_{\text{max}} = 1$, V must be $\pi/2$, and we can find the waveguide depth from Equation (10.34).

In the next section, we will apply Maxwell's equations to the planar dielectric waveguide.

10.2.2 Maxwell's Equations in the Planar Optical Waveguide

The optical waveguide has very low conductivity and so we can write $\sigma = 0$. Thus, Maxwell's equations become

$$\nabla \times \mathbf{E} = -\mu \frac{\partial \mathbf{H}}{\partial t} \tag{10.35}$$

and

$$\nabla \times \mathbf{H} = \varepsilon \frac{\partial \mathbf{E}}{\partial t} \tag{10.36}$$

Taking a transverse electric field propagating as in Figure 10.5, we get

$$\nabla \times \mathbf{E} = \begin{vmatrix} \mathbf{x} & \mathbf{y} & \mathbf{z} \\ \dfrac{\partial}{\partial x} & \dfrac{\partial}{\partial y} & \dfrac{\partial}{\partial z} \\ 0 & 0 & E_z \end{vmatrix} = \frac{\partial}{\partial y} E_z \mathbf{x} - \frac{\partial}{\partial x} E_z \mathbf{y} = -\mu \left(\frac{\partial}{\partial t} H_x \mathbf{x} + \frac{\partial}{\partial t} H_y \mathbf{y} \right) \tag{10.37}$$

and

$$\nabla \times \mathbf{H} = \begin{vmatrix} \mathbf{x} & \mathbf{y} & \mathbf{z} \\ \dfrac{\partial}{\partial x} & \dfrac{\partial}{\partial y} & \dfrac{\partial}{\partial z} \\ H_x & H_y & 0 \end{vmatrix} = \left(\frac{\partial}{\partial x} H_y - \frac{\partial}{\partial y} H_x \right) \mathbf{z}$$

$$= \varepsilon \frac{\partial}{\partial t} E_z \mathbf{z} \tag{10.38}$$

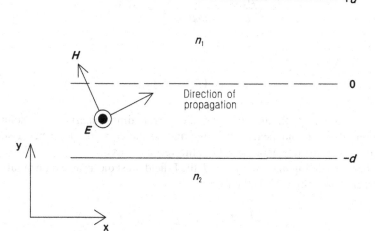

FIGURE 10.5 A transverse electric wave propagating in a planar dielectric waveguide. (Reprinted by permission from *Optical Communications*, Springer by Martin J. N. Sibley © 2019.)

The z component of the E field can be written as, in phasor notation,

$$\mathbf{E} = E\exp(j\omega t)\exp(-j\beta_x x)\mathbf{z}$$

with similar expressions for the two $\boldsymbol{H}$ field components. The propagation coefficient resolved along the x-axis is simply the phase coefficient β_x.

So, Equation (10.37) yields

$$\frac{\partial}{\partial y}E_z = -j\omega\mu H_x$$

and so

$$H_x = -\frac{1}{j\omega\mu}\frac{\partial}{\partial y}E_z \tag{10.39}$$

$$-j\beta_x E_z = j\omega\mu H_y$$

giving

$$H_y = -\frac{\beta_x}{\omega\mu}E_z \tag{10.40}$$

and

$$-j\beta_x H_y - \frac{\partial}{\partial y}H_x = j\omega\varepsilon E_z \tag{10.41}$$

Substitution of H_x and H_y into (10.41) gives

$$\frac{j\beta_x{}^2 E_z}{\omega\mu} + \frac{1}{j\omega\mu}\frac{\partial^2 E_z}{\partial y^2} = j\omega\varepsilon E_z$$

and so

$$\frac{\partial^2 E_z}{\partial y^2} = \left(\beta_x{}^2 - \omega^2\mu\varepsilon\right)E_z \tag{10.42}$$

which is valid in both the slab and the surrounding material. The solution to Equation (10.42) is independent of time and so the equation describes a standing wave pattern of the tangential E field in the vertical y-axis.

In the surrounding material, $y < -d$, the E field must die away exponentially as an evanescent wave. So, it takes the form

$$E_z = A\exp(\alpha_2 y) \quad y < -d \tag{10.43}$$

Substitution of Equation (10.43) into (10.42) yields

$$\alpha_2{}^2 = \beta_x{}^2 - \omega^2\mu\varepsilon$$

or

$$\alpha_2^2 = \beta_x^2 - n_2^2\beta_o^2 \tag{10.44}$$

where n_2 is the refractive index of the surrounding material and $\beta_o = \omega\sqrt{\mu_o\varepsilon_o}$ the free-space phase coefficient. By using Equation (10.39), the x-axis H field is given by

$$H_x = -\frac{1}{j\omega\mu}\frac{\partial}{\partial y} E_z$$

$$= \frac{-\alpha_2}{j\omega\mu} A\exp(\alpha_2 y) \quad y < -d \tag{10.45}$$

We need to find the constant A by using the continuity of the E and H fields at the boundary. For the fields inside the slab, we have

$$\frac{\partial^2 E_z}{\partial y^2} = \left(\beta_{1x}^{\;2} - n_1^2\beta_o^2\right)E_z$$

with a possible solution given by

$$E_z = B\sin(\beta_{1y}y + \phi) \quad -d < y < +d \tag{10.46}$$

where B is a constant, β_{1y} is the phase coefficient in the slab material resolved along the y-axis, and ϕ is a spatial phase shift yet to be determined. Thus, H_x will be given by Equation (10.39)

$$H_x = \frac{-B\beta_{1y}}{j\omega\mu}\cos(\beta_{1y}y + \phi) \quad -d < y < +d \tag{10.47}$$

If we take $y > d$, we are in the surrounding material and so must have an exponential decrease in E_z with increasing y. So, another solution to (10.42) is

$$E_z = C\exp(-\alpha_2 y) \quad y > d \tag{10.48}$$

and so

$$H_x = \frac{+\alpha_2}{j\omega\mu} C\exp(-\alpha_2 y) \quad y > d \tag{10.49}$$

where C is another constant. We now have six equations that describe the E and H fields within the slab and the surrounding material with various constants. We can now apply the boundary relationships to find these constants.

Let us consider the lower boundary first, i.e., $y = -d$. From Equations (10.43) and (10.46), we have

$$A\exp(-\alpha_2 d) = B\sin(-\beta_{1y}d + \phi) \quad \text{at } y = -d$$

and so

$$A = B\sin\left(-\beta_{1y}d + \phi\right)\exp(\alpha_2 d) \tag{10.50}$$

Using Equations (10.45) and (10.47), the continuity of the H_x field gives

$$\frac{-\alpha_2}{j\omega\mu} A\exp(-\alpha_2 d) = \frac{B\beta_{1y}}{j\omega\mu}\cos\left(-\beta_{1y}d + \phi\right)$$

and so

$$A = \frac{B\beta_{1y}}{\alpha_2}\cos\left(-\beta_{1y}d + \phi\right)\exp\left(\alpha_2 d\right) \tag{10.51}$$

Equating (10.50) and (10.51) gives

$$\tan\left(-\beta_{1y}d + \phi\right) = \frac{\beta_{1y}}{\alpha_2}$$

or

$$-\beta_{1y}d + \phi = \tan^{-1}\left(\frac{\beta_{1y}}{\alpha_2}\right) + m'\pi$$

where m' is an integer to take account of the periodicity of the tan function. Thus, the phase angle ϕ is given by

$$\phi = \tan^{-1}\left(\frac{\beta_{1y}}{\alpha_2}\right) + \beta_{1y}d + m'\pi \tag{10.52}$$

The continuity of the E field at the upper interface, $y = +d$ gives, using Equations (10.48) and (10.50),

$$C\exp(-\alpha_2 d) = B\sin\left(\beta_{1y}d + \phi\right)$$

which gives

$$C = B\sin\left(\beta_{1y}d + \phi\right)\exp\left(\alpha_2 d\right) \tag{10.52}$$

The continuity of the magnetic field, Equations (10.49) and (10.47) gives

$$\frac{\alpha_2}{j\omega\mu}C\exp(-\alpha_2 d) = -\frac{B\beta_{1y}}{j\omega\mu}\cos\left(\beta_{1y}d + \phi\right)$$

and so,

$$C = -\frac{B\beta_{1y}}{\alpha_2}\cos\left(\beta_{1y}d + \phi\right)\exp\left(\alpha_2 d\right) \tag{10.53}$$

By equating the two values for C, Equations (10.52) and (10.53), we get

$$B\sin\left(\beta_{1y}d+\phi\right)\exp\left(\alpha_2 d\right)=-\frac{B\beta_{1y}}{\alpha_2}\cos\left(\beta_{1y}d+\phi\right)\exp\left(\alpha_2 d\right)$$

Hence,

$$\tan\left(\beta_{1y}d+\phi\right)=-\frac{\beta_{1y}}{\alpha_2}$$

and so another expression for ϕ is

$$\phi=-\tan^{-1}\left(\frac{\beta_{1y}}{\alpha_2}\right)-\beta_{1y}d+m''\pi \tag{10.54}$$

where m'' is another integer. If we equate Equations (10.51) and (10.54), we get

$$\tan^{-1}\left(\frac{\beta_{1y}}{\alpha_2}\right)+\beta_{1y}d+m'\pi=-\tan^{-1}\left(\frac{\beta_{1y}}{\alpha_2}\right)-\beta_{1y}d+m''\pi$$

which gives

$$2\tan^{-1}\left(\frac{\beta_{1y}}{\alpha_2}\right)=-2\beta_{1y}d+m''\pi-m'\pi \tag{10.55}$$

The constants m' and m'' are both integers that cover 0 to ∞. We can arbitrarily set m' to zero. Now,

$$\tan^{-1}\left(\beta_{1y}/\alpha_2\right)=\frac{\pi}{2}-\tan^{-1}\left(\alpha_2/\beta_{1y}\right)$$

and so Equation (10.55) becomes

$$\frac{\pi}{2}-\tan^{-1}\left(\alpha_2/\beta_{1y}\right)=-\beta_{1y}d+\frac{m''\pi}{2}$$

and so,

$$\tan\left(\beta_{1y}d-N\frac{\pi}{2}\right)=-\frac{\alpha_2}{\beta_{1y}} \tag{10.56}$$

where N is an integer and the solution of this equation yields the values of β_{1y} for which light rays will propagate. This is identical to Equation (10.32) obtained using ray path analysis.

Let us now examine the cut-off condition for the planar waveguide. The maximum angle of incidence for any particular mode is the critical angle θ_c. If $\theta_i=\theta_c$, α_2 (Equation (10.31)) is zero and the E_z component is not attenuated as it passes through the surrounding material. Taking $\alpha_2=0$, $\beta_x=\beta_2$ and the phase coefficient of the propagating mode that of β is identical to that of the surrounding material.

This s known as the *cut-off condition,* and it represents the minimum value of β for which any mode will propagate. If the waveguide is operating well away from the cut-off condition, the propagating modes will be tightly bound to the slab, and so we can intuitively reason that $\beta_x = \beta_1$. Thus,

We can find the cut-off wavelength of any guide by substituting $\alpha_2 = 0$ into Equation (10.56) to give

$$\tan\left(\beta_{1y}d - N\frac{\pi}{2}\right) = 0$$

which implies

$$\beta_{1y}d = N\frac{\pi}{2} \tag{10.57}$$

As $\beta_{1y} = \beta_1\cos\theta_c$, $\cos^2\theta_c = 1 - \sin^2\theta_c$ and $\sin\theta_c = n_2/n_1$, we can write

$$\beta_1 d\sqrt{\left(1 - \frac{n_2^2}{n_1^2}\right)} = N\frac{\pi}{2}$$

or

$$\frac{2\pi n_1 d}{\lambda_{co}}\sqrt{\left(\frac{n_1^2 - n_2^2}{n_1^2}\right)} = N\frac{\pi}{2}$$

Thus, the cut-off wavelength for a particular mode is given by

$$\lambda_{co} = \frac{4d}{N}\sqrt{\left(n_1^2 - n_2^2\right)} \tag{10.58}$$

If we consider the lowest order mode ($N = 0$), we have a λ_{co} of infinity, and so there is, theoretically, no cut-off wavelength for the lowest order mode. If we take $N = 1$, we can determine the waveguide depth that just results in the first-order mode being cut-off.

Example

Light of wavelength 1.3 μm is propagating in a planar waveguide of width 200 μm and refractive index 1.46, surrounded by material of refractive index 1.44. Determine the waveguide depth for single-mode operation.

Let us consider the $N = 1$ mode and allow this mode to be just cut-off. The cut-off condition for this mode is

$$1.3\times 10^{-6} = 4d\sqrt{\left(1.46^2 - 1.44^2\right)}$$

and so the maximum waveguide depth is 2.7 μm.

Before we finish this particular section, let us return to the cut-off condition for any particular mode (10.58)

$$\beta_{1y} d = N \frac{\pi}{2}$$

This can be written as

$$V = N \frac{\pi}{2} \tag{10.59}$$

where

$$V = \frac{2\pi d}{\lambda_o} \sqrt{\left(n_1{}^2 - n_2{}^2\right)} \tag{10.60}$$

is known as the *V value* of the waveguide. We can also express V as

$$V^2 = \left(\alpha_2 d\right)^2 + \left(\beta_{1y} d\right)^2 \tag{10.61}$$

This can be easily proved by noting that

$$\alpha_2^2 = \beta_x^2 - n_2^2 \beta_o^2$$

and

$$\beta_{1y}^2 = n_1^2 \beta_o^2 - \beta_x^2$$

Thus,

$$V^2 = d^2 \left(\beta^2 - \beta_2^2 + \beta_1^2 - \beta^2\right)$$
$$= d^2 \left(n_1{}^2 \beta_o{}^2 - n_2{}^2 \beta_o{}^2\right)$$

and so,

$$V = \frac{2\pi d \sqrt{n_1{}^2 - n_2{}^2}}{\lambda_o}$$

which is identical to Equation (10.60).

In this section, we have applied Maxwell's equations to a planar dielectric waveguide. We finished by showing that we can reduce the number of propagating modes by decreasing the waveguide thickness. In particular, if the V value of the waveguide is less than $\pi/2$, then only the zero-order mode can propagate (so-called *single-mode* operation).

Problems

PROBLEMS FOR CHAPTER 1

1.1 Determine the circumference of a circle of radius r by integration. (Use the cylindrical coordinate set but ignore the z-axis. Take a small section of the circumference of length dl, subtending an angle dφ at the centre of the circle. It is then a matter of integrating with respect to φ between the limits 0 and 2π.)

Answer: $2\pi r$

1.2 Determine the area of a circle of radius r by integration. (Consider a small incremental area, ds. Take the thickness of this area to be dr, and let it subtend an angle of dφ to the centre of the disc. It is then a matter of integrating with respect to φ and then with respect to radius.)

Answer: πr^2

1.3 Determine the surface area of a sphere of radius r by integration. (Take a small incremental area, ds, on the surface of the sphere. Use the spherical coordinate set and integrate with respect to φ and Θ.)

Answer: $4\pi r^2$

1.4 Determine the volume of a cylinder of radius r and height l. (Use the result of Problem 1.2 and calculate the volume of a disc of height dz. Then integrate with respect to z.)

Answer: $\pi r^2 l$

1.5 Determine the volume of a sphere of radius r. (Problem 1.3 gave the surface area of a sphere. It is then a matter of calculating the volume of an incremental sphere of thickness dr, and integrating with respect to radius.

Answer: $\dfrac{4}{3}\pi r^3$

PROBLEMS FOR CHAPTER 2

2.1 Determine the flux radiating from a positive point charge of magnitude 400 pC. What is the flux density and electric field strength at a distance 10 mm from the charge? What is the force on a 10 µC point charge at this radius?

Answer: 400 pC; $0.32r$ µC m^{-2}; $36r$ kV m^{-1}; 0.36 N

2.2 A linear electric field of strength 10 V m^{-1} is established in free space. Determine the work done in moving a 10 µC charge a distance of 1 m against the field. Repeat if the field makes an angle of 30° to the direction of travel.

Answer: 0.1 mJ; 86.6 µJ

2.3   Determine the absolute potential at distances of 50 and 100 cm from a
      negative point charge of 20 µC. Hence find the potential difference between
      the outermost and innermost points.

Answer: −360 kV; −180 kV; 180 kV

2.4   Two charges, each of magnitude 10 pC, are situated on the x-axis at 1 and
      3 m from the origin. Determine the flux density, the electric field strength
      and the absolute potential at a point mid-way between them and 1 m along
      the z-axis. (Use the principle of superposition, i.e., consider each charge
      separately. When you are dealing with vectors, resolve them into horizontal
      and vertical components and then add components to get the resultant.)

Answer: $0.6z$ pC m^{-2}; $63.6z$ mV m^{-1}; 127 mV

2.5   Determine the flux density, electric field strength and absolute potential
      produced at a distance of 1 m from the centre of a 1 m radius disc with a
      charge of 10 µC evenly spread over it. Assume that the disc is in air.

Answer: $0.47z$ µC m^{-2}; 52.7 kV m^{-1}; 74.5 kV

2.6   A length of coaxial cable is to have a maximum capacitance of 100 pF
      m^{-1}. The cable is to be air-cored with a maximum electric field strength 3
      MV m^{-1}. As the cable is to be wound on a drum, the maximum outer radius
      is restricted to 3 cm. Determine the radius of the inner conductor and find
      the maximum voltage that can be carried.

Answer: 1.7 cm; 28.4 kV

2.7   A 200 m length of feeder consists of two 2-mm radius conductors separated
      by a distance of 20 cm in air. A potential of 1 kV is maintained between
      the two conductors. Determine the minimum and maximum values of field
      strength. If air breaks down at 3 MV m^{-1}, determine the maximum voltage
      that the feeder can withstand.

Answer: 55 kV m^{-1}; 2.2 kV m^{-1}; 55 kV

2.8   A 47 µF capacitor is charged to a potential of 50 V. The capacitor is then
      suddenly connected to a discharged 10 µF capacitor. Determine the energy
      stored in the 47 µF capacitor before and after the connection, and the energy
      stored in the 10 µF capacitor.

Answer: 58.8 mJ; 39.5 mJ; 8.4 mJ

PROBLEMS FOR CHAPTER 3

3.1   Determine the flux radiating from an isolated north pole of strength 400
      pWb. What is the flux density and magnetic field strength at a distance 10 cm
      from the pole? What is the force on a 10 µWb north pole at this radius?

Answer: 400 pWb; $3.2r$ nWb m^{-1}; $2.5r$ mN Wb^{-1}; $25r$ nN

3.2   A current element of length 1 µm passes a current of 1 A. This element
      is placed at the origin of a Cartesian coordinate set and is lying along the

vertical z-axis. Determine the magnetic field strength at a point whose coordinates are $x = 1$ m and $z -0.5$ m. Hence, find the force on a 10 µWb north pole placed at this point.

Answer: 57φ nN Wb^{-1}; 0.57φ pN

3.3 Find the flux density for the situation described in Problem 3.2. If the magnetic monopole is replaced by a current element of length 1 µm and current 3 A, determine the force on it. (Note the exceedingly small force on the current element. This is due to the $\boldsymbol{B}$ field being used to find the force on a current element.)

Answer: 7.2×10^{-14} φ Wb; 2.2×10^{-19} φ N

3.4 An ideal capacitor of value 100 pF is connected to an a.c. source of 10 V peak and frequency 200 Hz. Determine the current taken by the capacitor and hence the magnetic field strength at a distance of 5 cm from the capacitor leads. If you apply Ampere's law between the plates of the capacitor, is there still a field? If there is, where is the current enclosed by the path? (Use Ampere's law here. As regards the magnetic field between the plates, it is there and it is due to displacement current.)

Answer: 1.3 µA; 4.1 µWb

3.5 An air-spaced twin feeder transmission line carries a current of 1 kA. The conductors are spaced 5 m apart. Determine the force per unit length between the conductors.

Answer: 0.04 N m^{-1}

3.6 Determine the inductance per unit length for the coaxial cable designed in Problem 2.6.

Answer: 164 nH m^{-1}

3.7 A length of 4-mm wide microstrip is etched on one side of some double-sided pcb. The thickness of the board is 3 mm and the non-ferrous dielectric has a relative permittivity of 3. Estimate the capacitance and inductance per cm.

Answer: 0.35 pF cm^{-1}; 9.9 nH cm^{-1}

3.8 An alternative method of finding the internal inductance of a piece of wire is one based on energy storage. This equates the stored energy in the wire, obtained from a 'fields' point of view, to that obtained from a 'circuits' point of view. Use this method to obtain the internal inductance of a piece of wire. (Refer to Figure 3.14a and find the magnetic field at radius r – see Example 3.2 in Section 3.6. Then find the energy per m^3 from Equation (3.67) and hence find the fractional stored energy in the tube of Figure 3.14a. It is then a case of integrating with respect to radius to obtain the total energy stored in the wire. This can be equated to that obtained for Equation (3.66) to obtain the inductance.)

PROBLEMS FOR CHAPTER 4

4.1   The mobility of an electron in a metal is given by

$$\mu = \frac{q\tau}{2m}$$

where τ is the mean time between collisions and m is the mass of an electron (9.11×10^{-31} kg). Determine the time between collisions for an electron in copper with 8.5×10^{28} free electrons per m^3 and $\sigma = 58$ MS m^{-1}.

Answer: 4.9×10^{-14} s

4.2   If the sample in Problem 4.1 has a current density of 1×10^6 A m^{-2}, determine the electric field strength. Hence, estimate the drift velocity of the electrons.

Answer: 17.2 mV m^{-1}; 73.5 μm s^{-1}

4.3   A 1-m long cylindrical block of metal has a resistance of 5 μΩ and radius 5 cm. The block is drawn through a die into wire of radius 1 mm. Determine the resistance per metre length of wire.

Answer: 12.5 mΩ

4.4   A length of coaxial cable has an inner conductor of radius 2 mm and an outer conductor of radius 2 cm. The conductivity of the dielectric is 5×10^{-4} S m^{-1} and it has a relative permittivity of 4. The cable carries a voltage of 1 kV at a frequency of 1 MHz. Determine the resistance of a 100 m length and hence find the loss tangent. What power is lost due to the shunt resistance?

Answer: 7.33 Ω; 2.25; 136 kW

PROBLEMS FOR CHAPTER 6

6.1   Show that the absolute potential at a point P distance R from an electric dipole is given by

$$V = \frac{p\cos\theta}{4\pi\varepsilon R^2}$$

where θ is the angle made by a line joining P to the dipole. (Draw the dipole pointing upwards with the point P to the right-hand side. The angle θ is the internal angle. Also, use the principle of superposition.)

6.2   A 1 V m^{-1} electric field, travelling in air, is obliquely incident on a block of glass, $\varepsilon_r = 6$. Determine the angle the transmitted field makes to the boundary if the field makes an angle of 45° to the glass block.

Answer: 9.5°

PROBLEMS FOR CHAPTER 7

7.1 A hysteresis loop has an area of $100\,cm^2$ when drawn with $1\,cm$ representing 25 At m^{-1} and $1\,cm$ representing 0.2 Wb m^{-2}. Determine the hysteresis loss per unit volume if the sample is used in a transformer operating at 50 Hz.

Answer: $25\,kW\ m^{-3}$

7.2 The bobbin of a moving-coil meter has 40.5 turns of wire wound on it and is placed in a magnetic field which has a flux density of 20 mWb m^{-2}. The bobbin is connected to an indicating mechanism with a restoring spring that exerts a torque of 1 μN m per degree of deflection. The coil is 5 cm long and has a diameter of 3 cm. Determine the current required to give a deflection of 15°.

Answer: 12.35 mA

7.3 Repeat the example of Section 7.6 using the B/H data given below:

B mWb m^{-2}	30	400	430	80	110
H kAt m^{-1}	1.4	5.2	6.0	2.6	3.1

Answer: 8.6 A; 4.15 A

PROBLEMS FOR CHAPTER 8

8.1 A length of coaxial cable has an inductance of 160 nH m^{-1} and a capacitance of 100 pF m^{-1}. Determine, at a frequency of 1 MHz, the phase shift per unit length; the phase velocity; the time taken for a signal to travel 100 m; the wavelength along the line.

Answer: $0.025\,rad\ m^{-1}$; $2.5 \times 10^8\,m\ s^{-1}$; 400 ns; 250 m

8.2 A p.c.b. track is 5 cm long and has the parameters of Problem 3.7. Determine the length of time a signal takes to reach the end. Compare this to the pulse time of a 1 Gbit s^{-1} pulse. (A 1 Gbit s^{-1} pulse is 1 ns long.)

Answer: 0.3 ns

8.3 A lossy line has the following parameters:

$$R = 10\ \Omega m^{-1}\ L = 150\ nH\ m^{-1}$$
$$G = 100\ \mu S\ m^{-1}\ C = 100\ nF\ m^{-1}$$

Determine the propagation coefficient, the attenuation coefficient and the phase coefficient at a frequency of 500 kHz. Compare these values to those for a frequency of 10 MHz. Comment on the result.

Answer: $0.032\underline{/1.3}$; $0.032\,m^{-1}$; $726 \times 10^{-6}\,rad\ m^{-1}$; $0.037\underline{/25}$; $0.034\,m^{-1}$; $0.016\,rad\ m^{-1}$. The reactive components are dominant and the line becomes lossless.

8.4 The load on a transmission line becomes open-circuit due to a fault. Determine the reflected and load voltage. Repeat if the fault in the load is a short-circuit.

Answer: V_i; $2V_i$; $-V_i$; 0

8.5 In very high-speed logic systems, the load must be matched to the transmission line impedance to stop reflections. Confirm that the reflection coefficient is zero.

8.6 An octal buffer device switches all eight outputs at the same time. This causes a transient to occur on the 5 V power rail as the current drawn suddenly becomes 80 mA. The power rail is narrow and has a characteristic impedance of 100 Ω. Determine the magnitude of the voltage pulse that travels to the system power supply. The effect of this transient can be eliminated by using a decoupling capacitor. Where should this be situated?

Answer: 8 V; as close to the power pin on the i.c. as possible.

PROBLEMS FOR CHAPTER 9

9.1 Calculate the propagation coefficient for glass ($\varepsilon_r = 2$, $\mu_r = 1$, $\sigma = 1 \times 10^{-4}$ S m⁻¹) at frequencies of 10 MHz and 100 THz.

Answer: $0.3\underline{/-2.5°}$, $j3 \times 10^6$

9.2 Calculate, at a frequency of 1 MHz, the skin depth of tin ($\sigma = 9.2 \times 10^6$ S m⁻¹), copper ($\sigma = 58 \times 10^6$ S m⁻¹) and aluminium ($\sigma = 37 \times 10^6$ S m⁻¹). Hence, determine the best material for electromagnetic shielding.

Answer: 0.17 mm, 66 μm, 83 μm, copper

Bibliography

Basic Electromagnetism and its Applications Compton, A J, Chapman and Hall, 1986, ISBN 978-0-412-38130-0.

Electricity and Magnetism, 3rd ed. Purcell, E M and Morin, D J, Cambridge University Press, 2013, ISBN 978-1-107-01402-2.

Electromagnetism, 2nd ed. Grant, I S and Phillips, W R, Wiley, 1990, ISBN 0 471 92711 2.

Electromagnetism for Engineers: An Introductory Course, 3rd ed. Hammond, P, Pergamon Press, 1986, ISBN 0-08-032583-1.

Electromagnetism: Problems with Solutions, 3rd ed. Pramanik, A, PHI Learning, 2012, ISBN 978-81-203-4633-8.

Introduction to Electrodynamics, 4th ed. Griffiths, D J, Cambridge University Press, 2019, ISBN 978-1-108-42041-9.

Solved Problems in Classical Electromagnetism: Analytical and Numerical Solutions with Comments Pierrus, J, Oxford University Press, 2018, ISBN 978-0-19-882192-2.

Index

Printed in the United States
By Bookmasters